RESILIENCE THROUGH KNOWLEDGE CO-PRODUCTION

Confronted with the complex environmental crises of the Anthropocene, scientists have turned to interdisciplinarity to grapple with challenges that are at once social and ecological. Indigenous knowledge holders have contributed critical observations and understandings and, in doing so, have gained global recognition. Most recently, several arenas are calling for the co-production of new knowledge by bringing together Indigenous knowledge and science. This new call to arms is rapidly gaining momentum, but little guidance has been offered on how co-production can be achieved.

This book revisits epistemological debates on the notion of co-production, and assesses methods, principles and values that have allowed "communities of practice" involving Indigenous experts and scientists to jointly co-produce knowledge. This challenging undertaking calls into question many of our assumptions about the breadth and limits of knowledge, both Indigenous and scientific, about the interactions between environment and society. In view of their distinct ontologies and epistemologies, their synergies but also incompatibilities, as well as persistent asymmetries of power, what are the determining factors for achieving an Indigenous-scientific knowledge co-production rooted in equity, mutual respect and shared benefits?

Resilience through Knowledge Co-Production includes several collective papers co-authored by Indigenous knowledge holders and scientists, with case studies involving Indigenous communities from the Arctic, Pacific islands, the Amazon, the Sahel and high-altitude zones. This book offers guidance to Indigenous peoples, scientists, decision-makers and NGOs on how to move toward a decolonized co-production of knowledge that brings together Indigenous knowledge and science.

MARIE ROUÉ is Emerita Director of Research at the National Centre for Scientific Research and the National Museum of Natural History in France. From 1995 to 2000, she was Director of the Apsonat (Appropriation and Socialization of Nature) research team. She has published several books as well as numerous papers and book chapters about Indigenous peoples in the Arctic and Subarctic, and edited special journal issues on biodiversity and cultural diversity, ecological utopia, human–animal relationships, and NGOs & Indigenous knowledge. Her field research focuses on Indigenous knowledge and global change among the Sámi (Norway and Sweden), Cree First Nations and Inuit (Canada) and rural communities in the Cevennes (France). She served as a member of the Multidisciplinary Expert Panel (MEP) and Indigenous and Local Knowledge Task Force of the Intergovernmental Platform on Biodiversity and Ecosystem Services (IPBES).

DOUGLAS NAKASHIMA has been working in the field of Indigenous knowledge for over thirty-five years, with his initial research focusing on Inuit and Cree First Nations in Arctic and subarctic Canada. He recently retired from UNESCO, where he worked in the Natural Sciences Sector from 1996 to 2018. In 2002, he founded UNESCO's global programme on Local and Indigenous Knowledge Systems (LINKS) that addresses the role of Indigenous knowledge in environmental management, including in response to climate change and biodiversity loss. Dr Nakashima led UNESCO's work with the IPCC to highlight the importance of Indigenous knowledge for climate change assessment and adaptation, including publication of the compendium *Weathering Uncertainty: Traditional Knowledge for Climate Change Assessment and Adaptation*.

IGOR KRUPNIK is Curator of the Arctic Ethnology collections at the National Museum of Natural History, Smithsonian Institution in Washington, DC. Trained as a cultural anthropologist and ecologist, Dr Krupnik has worked among the Yupik, Chukchi, Aleut, Nenets and Inupiaq peoples, primarily in Alaska and the Russian Arctic region. His area of expertise lies in modern cultures, Indigenous ecological knowledge, and the impact of modern environmental and social change on human life in the North. He has published more than twenty books, catalogues, and edited collections, including several "sourcebooks" on Indigenous ecological and historical knowledge produced jointly with local partners for community use.

RESILIENCE THROUGH KNOWLEDGE CO-PRODUCTION

Indigenous Knowledge, Science and Global Environmental Change

Edited by

MARIE ROUÉ
National Centre for Scientific Research (CNRS)

DOUGLAS NAKASHIMA
UNESCO

IGOR KRUPNIK
Smithsonian Institution

This book should be cited as

Roué, M., Nakashima, D. and Krupnik, I. 2022. *Resilience through Knowledge Co-production: Indigenous Knowledge, Science and Global Environmental Change*. Local & Indigenous Knowledge 3. Cambridge University Press and UNESCO: Cambridge and Paris.

University Printing House, Cambridge CB2 8BS, United Kingdom

One Liberty Plaza, 20th Floor, New York, NY 10006, USA

477 Williamstown Road, Port Melbourne, VIC 3207, Australia

314–321, 3rd Floor, Plot 3, Splendor Forum, Jasola District Centre, New Delhi – 110025, India

103 Penang Road, #05-06/07, Visioncrest Commercial, Singapore 238467

Cambridge University Press is part of the University of Cambridge.

It furthers the University's mission by disseminating knowledge in the pursuit of education, learning, and research at the highest international levels of excellence.

Published jointly by the United Nations Educational, Scientific and Cultural Organization (UNESCO), 7, Place de Fontenoy, 75007 Paris, France, and Cambridge University Press, University Printing House, Shaftesbury Road, Cambridge CB2 8BS, United Kingdom.

www.cambridge.org
Information on this title: www.cambridge.org/9781108838306
DOI: 10.1017/9781108974349

© UNESCO 2022
First published 2022

Printed in the United Kingdom by TJ Books Limited, Padstow Cornwall

A catalogue record for this publication is available from the British Library.

Library of Congress Cataloging-in-Publication Data
Names: Roué, Marie, editor. | Nakashima, D. J., editor. | Krupnik, Igor, editor.
Title: Resilience through knowledge co-production : indigenous knowledge, science, and global environmental change / edited by Marie Roué, Douglas Nakashima, Igor Krupnik.
Description: Cambridge ; New York, NY : Cambridge University Press, 2022. | Includes bibliographical references and index.
Identifiers: LCCN 2021058058 (print) | LCCN 2021058059 (ebook) | ISBN 9781108838306 (hardback) | ISBN 9781108974349 (epub)
Subjects: LCSH: Indigenous peoples–Ecology–Case studies. | Traditional ecological knowledge–Case studies. | Ethnoecology–Case studies. | Human beings–Effect of climate on. | Climatic changes. | BISAC: SCIENCE / Earth Sciences / Meteorology & Climatology
Classification: LCC GN476.7 .R47 2022 (print) | LCC GN476.7 (ebook) | DDC 306.4/5–dc23/eng/20220110
LC record available at https://lccn.loc.gov/2021058058
LC ebook record available at https://lccn.loc.gov/2021058059

ISBN 978-1-108-83830-6 Hardback
UNESCO ISBN: 978-92-3-100516-9

Contents

Colour Plates section to be found between pp. 142 and 143

Contributors

Julia Vieira da Cunha Ávila, Research Group in Archaeology and Amazonian Cultural Heritage, Mamirauá Institute for Sustainable Development, Tefé, Brazil; and Graduate Studies Programme in Botany, National Institute for Amazonian Research, Manaus, Brazil.

Annett Bartsch, b.geos, Korneuburg, Austria; and Austrian Polar Research Institute, Vienna, Austria.

Linette N. Boisvert, Earth System Science Interdisciplinary Center, University of Maryland, College Park, Maryland, USA.

Anders Henriksen Bongo is a Sámi reindeer herder from northern Norway. After enduring the vicissitudes of herding for several decades, and confronted with health problems, he finally took the decision to stop herding and settle permanently in Kautokeino, the main winter village of the herding community. He went on to work for the reindeer-herding administration before becoming the head of a company fabricating tents inspired by the traditional Sámi design. Highly reputed as a herder, he never ceased to observe, analyse and monitor the evolution of reindeer herding and its challenges. Today, at eighty-two years of age, he provides observations and reflections on the changes he has observed throughout the course of his life.

Anders Burman received his PhD in social anthropology from the University of Gothenburg in 2009. He was a postdoctoral scholar at the Department of Ethnic Studies at the University of California–Berkeley and is currently a senior lecturer at the Human Ecology Division at Lund University. He has published on Indigenous peoples and movements, activism, cosmology, gender politics, political ontology, decolonization and knowledge production, in the fields of political ecology and environmental anthropology, with a geographical focus on the Andes, Bolivia and Latin America. Currently engaged in a research project focusing on how climate change is perceived and explained differently by different actors from different ontological life worlds in Latin America, he focuses on the contradictions arising from the encounter between hegemonic notions of "nature," "climate," Indigenous knowledge, the Andean landscape and the cosmos.

Mar Cabeza is a lecturer in conservation science at the Faculty of Biological and Environmental Sciences of the University of Helsinki (Finland) and the Principal Investigator of the Global Change and Conservation lab. Her most significant research has focused on addressing climate change impacts on biodiversity and their contribution to paradigm shifts in conservation approaches. She currently works on a broad range of research topics related to conservation planning and assessments, offering an integrated socioecological approach to conservation issues.

Hanafi Amirou Dicko, a traditional herder from Burkina Faso, is the president of the Association of Traditional Breeders of Sahel (Association des Eleveurs Traditionnels du Sahel, or Dawla Sahel). This association is involved in the protection of traditional breeders' rights within national development policies and in the strengthening of Sahel breeders' resilience in the face of climate change. He has collaborated with international teams, including the Meteorology General Direction of Burkina Faso, in the context of pluridisciplinary research on communication of seasonal rainfall forecasts with farmers and agro-pastoralists. El Hadji Dicko is a member of the executive committee of IPACC (Indigenous Peoples of Africa Co-ordinating Committee), which is a network of 135 associations of Indigenous peoples in twenty-one African countries. A retired technical agent from the Ministry of Animal Resources of Burkina Faso, he is also a representative of the Federation of Burkina Breeders of Sahel (Fédération des Eleveurs du Burkina).

Matthew L. Druckenmiller is a research scientist with the National Snow and Ice Data Center (NSIDC) at the University of Colorado–Boulder. Since 2006, Matthew has worked within the coastal regions of Arctic Alaska, investigating the connections between changing sea ice conditions and marine mammal habitat, and local Indigenous community use of sea ice for hunting and travel. As of 2021, he serves as Director of the Navigating the New Arctic Community Office (NNA-CO) and co-leads the Exchange for Local Observations and Knowledge of the Arctic (ELOKA). Matthew also serves as the lead US delegate to the International

Arctic Science Committee (IASC), an editor for the National Oceanic and Atmospheric Administration (NOAA) Arctic Report Card, and an editor for the Arctic Chapter within the *Bulletin of the American Meteorological Society*'s (BAMS) annual *State of the Climate* report. Matthew earned his PhD in geophysics from the University of Alaska–Fairbanks in 2011.

Hajo Eicken is Director of the International Arctic Research Center and a professor of geophysics at the University of Alaska–Fairbanks, where he specializes in the field of sea ice geophysics. His research includes studies of the growth, evolution and properties of sea ice in the Arctic and Antarctic. He is particularly interested in determining how small-scale properties and the microstructures of sea ice impact processes on a larger scale, from ecosystems to the climate system. Eicken studies the multiple uses of sea ice; the services it provides – from cooling the planet to serving as a platform for animals and people; and its broader relevance in planning for Arctic futures.

Álvaro Fernández-Llamazares is a postdoctoral fellow at the Helsinki Institute of Sustainability Science (HELSUS) of the University of Helsinki, Finland. During his PhD studies, he conducted research on the effects of global change on the biocultural heritage of the Tsimane' people of Bolivian Amazonia. His postdoctoral research investigates biocultural conservation approaches in Kenya, Bolivia and Madagascar, focusing on how to connect global policy discourses around conservation with the on-the-ground realities of Indigenous peoples and local communities.

Bruce C. Forbes (PhD geography, 1993, McGill University) leads the Global Change Research Group at the Arctic Centre, University of Lapland in Rovaniemi, Finland. He is also a docent in plant ecology/biogeography at the Faculty of Science, University of Oulu, Finland. He has a background in applied ecology and geography in northern high latitudes, with special emphasis on permafrost regions. His experience is circumpolar, encompassing studies of rapid land-use and climate change in Alaska, the Canadian High Arctic, various regions of northern Russia and northernmost Fennoscandia. His approach is strongly interdisciplinary and participatory, aiming for the co-production of knowledge, particularly concerning local and regional Indigenous stakeholder-driven research questions. He has conducted fieldwork annually in the Arctic for the past thirty-six years. For the past twenty years, his research has focused on resilience in social-ecological systems in close cooperation with Indigenous Nenets and Sámi reindeer herders; analyses of proxy data sources for climate change, including extreme weather events; and vegetation productivity in northwest Eurasian tundra ecosystems.

Tikoidelaimakotu Tuimoce Fuluna, from Moce island in the Lau Group, Fiji, lives on land in the capital city of Fiji that was occupied by his father in the early 1990s. His father and grandfather sailed *druas* (traditional canoes) to the city to start a traditional canoe cruise business. On his fourth trip, his dad lost his life, and after that Fuluna was raised by his mother. This land has become home to his relatives, who continue to keep their tradition alive two decades later through traditional canoe building and *masi* making. Living in the city and practicing their tradition help keep their identity and Indigenous knowledge alive. They do not rely on engines to gather their food, but rather use the wind to power their traditional canoes for fishing or gathering seashells. In 2010, Fuluna had the good fortune to voyage halfway around the world on a traditional canoe powered by wind and solar energy to spread the message that we do not need to burn fossil fuels to travel the world. This trip reminded him of how his forefathers used to sail from island to island and underlined the importance of Indigenous knowledge to fight climate change.

Henry P. Huntington is an independent researcher living in Eagle River, Alaska. Most of his work examines human–environment interactions in the Arctic, especially involving Indigenous peoples. Henry has worked in Alaska, Canada, Russia, Nepal and elsewhere, and has contributed to several major international Arctic assessments. He served as co-chair of a US National Academy of Sciences committee on emerging research questions in the Arctic. He is also the Arctic science director for Ocean Conservancy, working to conserve the Arctic Ocean and surrounding seas.

Kwang-Yul Kim, Earth and Environmental Sciences, Seoul National University, Seoul, Republic of Korea.

Igor Krupnik (see co-editor bionote on page ii)

Timo Kumpula, Geographical and Historical Studies, University of Eastern Finland, Joensuu, Finland.

Roza Laptander, Arctic Centre, University of Lapland, Rovaniemi, Finland.

Marc Macias-Fauria, School of Geography and the Environment, University of Oxford, Oxford, United Kingdom.

Nina Meschtyb, Arctic Centre, University of Lapland, Rovaniemi, Finland.

Douglas Nakashima (see co-editor bionote on page ii)

George Noongwook is a whaling captain, drummer, dancer and scholar from Savoonga, Alaska. He has served as Chairman and Vice-Chairman of the Alaska Eskimo Whaling Commission, representing the interests of Alaska's whaling communities nationally and internationally. George has also led several research projects in Savoonga, a St. Lawrence Island Yupik community of about 700 people in the northern Bering Sea.

Lars-Evert Nutti is a Sámi herder from Sirges, a community of reindeer herders (Swedish *sameby*) in Jokkmokk, northern Sweden. In addition to his herding activities, he is actively involved in the dialogue between the forestry sector and reindeer husbandry, engaging regularly in dialogues at the national level. He has become an expert in forestry–reindeer husbandry interactions in Sápmi.

Hindou Oumarou Ibrahim is a member of the Mbororo (Peulh) community of Chad. She is an Indigenous expert on climate change adaptation and mitigation, traditional knowledge and pastoralism in Africa; a founding member of the Marrakesh Platform for Climate Action; and Co-chair of the International Indigenous Peoples' Forum on Climate Change. She is also a member of the scientific and technical committee for Biosphere and Heritage of Lake Chad (BIOPALT) of UNESCO, as well as a steering group member of the Executive Committee of IPACC (Indigenous Peoples of Africa Coordinating Committee), where she serves as the focal point for climate change adaptation, traditional knowledge and gender. She was awarded an Emerging Explorer Prize by *National Geographic* in 2017, as well as the Prix Spécial Danielle Mitterrand, and was recognized as one of the 100 most influential women in the world by the BBC in 2018. On three occasions, she addressed the United Nations Security Council on climate change and international insecurity.

Jacson Rodrigues, Agroecosystems Management Program, Mamirauá Institute for Sustainable Development, Tefé, Brazil.

Camille Rognant, Laboratoire Environnement Villes Sociétés, Université Lumière Lyon 2, Lyon, France.

Samuel Roturier is an assistant professor at AgroParisTech in the Laboratoire Ecology, Systematic & Evolution, Paris-Saclay University, France, and has a doctoral degree in biology and environmental anthropology. He has been working for more than fifteen years in Swedish Sápmi in collaboration with Sámi reindeer-herding communities and forest companies. He is an expert in Sámi ecological knowledge and interactions with the boreal forest.

Marie Roué (see co-editor bionote on page i)

Jan Salick is Senior Curator Emerita at the Missouri Botanical Garden, David Fairchild Medalist for Plant Exploration and Distinguished Economic Botanist. She researches and publishes on ethnobotany and ethnoecology. She, her collaborators and her students have been studying the devastating impacts of environmental changes on Indigenous peoples around the world. Their work on climate change dates back more than twenty years and is recognized internationally among scientists and policymakers, especially the Millennium Ecosystem Assessment, Intergovernmental Panel on Climate Change (IPCC), Global Partnership for Plant Conservation (GPPC) and Intergovernmental Platform on Biodiversity and Ecosystem Services (IPBES). Dr. Salick has received funding from the US National Science Foundation, the National Institutes of Health and National Cancer Institute (NIH and NCI), the Ford Foundation, The Nature Conservancy, the National Geographic Society and other sources. From the tropical rainforests of Peru and Borneo and the high alpine ecosystem of the Himalayas, to her most recent work with American Indians, she applies her scientific results to support Indigenous peoples and to raise awareness of the impacts of environmental and climate change around the world.

Anne K. Salomon is a marine ecologist and associate professor in the School of Resource and Environmental Management at Simon Fraser University in British Columbia, Canada. Anne is interested in the cascading effects of predators on marine food webs, marine reserve design and evaluation, regime shifts and tipping points in coastal ocean ecosystems, and the resilience of social-ecological systems.

Jessica Poliane Gomes dos Santos, Research Group in Geospatial Analysis and Amazonian Territories, Mamirauá Institute for Sustainable Development, Tefé, Brazil; and Graduate Studies Program in Geography, Federal University of Minas Gerais, Belo Horizonte, Brazil.

Rafael Barbi Costa e Santos, Graduate Studies Program in Anthropology, University of Brasília, Brasília, Brazil.

Anna Skarin, Swedish University of Agricultural Sciences, Uppsala, Sweden.

Angela May Steward (PhD in 2008 from the City University of New York and New York Botanical Gardens) is a research scientist who has been studying livelihood transformations in traditional Amazonian communities (Indigenous, *ribeirinho* and Quilombo) since 2004. Her expertise is in ethnobiology and environmental anthropology, with a particular interest in understanding practice, knowledge and change in traditional agricultural systems. Her geographic focus is Brazil, particularly the Brazilian Amazon, but she has also worked in savanna (*cerrado*) environments. From 2011 to 2016, she worked in the middle Solimões region in Central Amazonia as a researcher and program coordinator of the Agroecosystems Management Program at the Mamirauá Institute for Sustainable Development, focusing on how floodplain producers are adapting their strategies and production systems in the face of changing flood patterns. Since 2016, she is an associate professor of agrarian studies and rural development at the Federal University of Pará, Belém, Brazil.

Julienne C. Stroeve, National Snow and Ice Data Center, University of Colorado–Boulder, USA, and University College London, UK.

Nick M. Tanape, Sr., was a Sugpiaq Elder from Nanwalek, Alaska, and a Native community representative at the Pratt Museum in Homer, Alaska. An artist, fisherman and hunter, Nick led the construction of traditional boats in his community, restoring skills in craftsmanship and seamanship. He was well known for his depth of knowledge about Sugpiaq culture, hunting practices and the ocean environment of the lower Kenai Peninsula and Gulf of Alaska. Sadly, Nick passed away in 2018 at the age of seventy-two.

Nils-Johan Utsi is a Sámi herder from Sirges, a community of reindeer herders (Swedish *sameby*) in Jokkmokk, northern Sweden. He has always demonstrated an interest in exchanging with scientists.

Mariana Verdonen, Geographical and Historical Studies, University of Eastern Finland, Joensuu, Finland.

Fernanda Maria de Freitas Viana, Agroecosystems Management Program, Mamirauá Institute for Sustainable Development, Tefé, Brazil.

Samis Vieira, Agroecosystems Management Program, Mamirauá Institute for Sustainable Development, Tefé, Brazil.

Winton Weyapuk, Jr., was born in Wales, Alaska, in 1950. He was a subsistence hunter from a young age, became a whaling captain, and served as Chair of the Wales Whaling Captains Association. He received a BA in rural development with an emphasis in land use planning in 1986 and a BA in Iñupiaq Eskimo language from the University of Alaska–Fairbanks in 1987. He served on the Wales Native

Corporation's Board of Directors and on the Native Village of Wales (IRA) Council and held many other roles in his community. He generously shared his knowledge with students at the Kiŋikmiut School in Wales as well as collaborated with researchers throughout the country. As part of these collaborations, Mr. Weyapuk carried out a decade of sea ice observations and was instrumental in publishing the *Wales Iñupiaq Sea Ice Dictionary*. He passed away in 2016.

Hans Winsa has worked for Sveaskog, the Swedish national forest company, for more than twenty years. He completed a PhD in silviculture in 1995. During his career, he worked as a forest district manager in Norrbotten and then at the Head Office, retiring in 2018.

Pentti Zetterberg, Dendrochronology Laboratory, University of Eastern Finland, Joensuu, Finland.

Acknowledgements

This book would not exist without the generous contributions of Indigenous peoples who have shared their knowledge and experiences as volume authors or project partners. Many chapters in this book began as presentations by Indigenous knowledge holders, natural and social scientists, and governmental and nongovernmental actors at international events, including:

the international Global Change in the Arctic and Co-production of Knowledge workshop, 27–29 September 2012, National Museum of Natural History, Paris, France;

the international Resilience in a Time of Uncertainty: Indigenous Peoples and Climate Change conference at UNFCCC COP 21, 26–27 November 2015, UNESCO, Paris, France; and

the international Indigenous Knowledge and Climate Change conference at UNFCCC COP 22, 2–3 November 2016, Marrakech, Morocco.

Special thanks to H. Oumarou Ibrahim and the International Indigenous Peoples' Forum on Climate Change for their active participation. We are grateful for support provided by G. Boeuf, former president of the National Museum of Natural History, France; P. Monfray and the Agence Nationale de la Recherche (ANR), France; as well as O. Fontan and the Ministry of Foreign Affairs and International Relations, France. Sincere thanks to the late L. Kielsen-Holm and S. Gearheard for contributing to debates on Arctic climate change.

Event organization was enabled by T. Lahrem at the National Museum of Natural History, H. Belguenani and the UNESCO Office in Rabat, and the UNESCO-LINKS team in Paris, including S. Fadina, C. Hauke, J. Cheftel, S. Cogos, V. Gonzalez-Gonzalez, S. Kang, T. Renard–Truong Van Nga and A. Rodriguez. Special thanks to J. Rubis for leading the LINKS-Climate Frontlines project and N. Crawhall for perpetuating the UNESCO-LINKS programme.

J. M. Bongo, A. Dicko and G. Mitchel provided generous and able assistance with the Indigenous contributions to the volume. J. Maréchal helped compile papers and provided transcriptions and translations. T. Narayan provided English editing and indexing for the volume. Special thanks to R. Swinnen, I. Denison and C. Puerta at UNESCO Publishing and E. Kiddle, S. Lambert, S. Duveau, N. Harikrishnan and K. Hughes at Cambridge University Press for their patient guidance throughout the production of the book.

Introduction

1

Co-production between Indigenous Knowledge and Science: Introducing a Decolonized Approach

MARIE ROUÉ AND DOUGLAS NAKASHIMA

Times are bleak for planet Earth. The contours of the challenges facing human society are more and more sharply defined: global climate change, unprecedented biodiversity loss, deepening poverty and growing inequalities. We have entered the era of the Anthropocene. Our society's capacity to transform the environment in which we live has become a threat to our own existence. This new global outlook contrasts sharply with the twentieth-century myth of human progress orchestrated by advances in science and technology. Our heavy hand on the planet now confronts humans with a future full of unknowns.

This threat to human well-being is further aggravated by climate injustice. Societies transformed by scientific and technological 'progress' are consuming by far the greatest proportion of the earth's resources and are responsible for the vast majority of greenhouse gas emissions. Yet, they rarely bear the brunt of negative repercussions. Among those suffering the most are Indigenous peoples and rural communities who continue to rely on natural environments that are increasingly under threat. Pacific islanders are combatting rising sea levels that exacerbate coastal erosion and decimate food crops as a result of saltwater intrusion. The lives and livelihoods of circumpolar Arctic peoples are threatened by regional warming at rates twice the global average. Herders and small-scale farmers are facing increasingly frequent and extreme drought and flooding throughout the African Sahel.

Nevertheless, Indigenous and rural communities around the world are not passive victims of global change. They are demonstrating an astounding capacity to innovate and adapt. This resilience is rooted in in-depth intergenerational knowledge and know-how about the natural milieu and its variability, as well as social systems and community values that ensure solidarity and the sharing of lessons learned.

As our impacts on the planet move us into uncharted waters, scientists have had to take a step back from their habitual 'expert' stance. As science is no longer able to offer certitudes about the future state of the earth, opportunities have been created to recognize the observations and understandings that other knowledge systems bring to the decision-making table. This fracturing of science's hegemony over 'truth' has facilitated the formation of collaborative partnerships across knowledge systems that enhance our understanding of a world in rapid transformation. One such partnership, the decolonized knowledge co-production between Indigenous knowledge holders and scientists (DKC), is the subject of this book.

In this introductory chapter, we outline our proposal. Our usage of the term 'co-production' differs from much of the contemporary scientific literature, which focuses on co-production within Western society, either across scientific disciplines or among scientists, citizens and policymakers (Meadow et al., 2015; Lemos et al., 2018; Miller and Wyborn, 2018; Adelle et al., 2020). Our endeavour explicitly reaches beyond Western society to co-produce knowledge at the interface of science and Indigenous knowledge systems. Knowledge co-production is not, however, just another way to refer to traditional or Indigenous knowledge. Instead, we propose Indigenous–scientific knowledge co-production as an emerging practice that has as yet rarely been attained in full, a challenging goal towards which future collaborations between Indigenous knowledge holders and scientists should aspire.

This chapter is organized into four sections, with the first three relating to the origins of our concept of co-production, its theoretical foundations and essential constituent parts, followed by a fourth section that offers an overview of the book. Descriptions of each section follow:

1 *Origins: Ethnoscience and Indigenous Knowledge of the Environment*

Our conception of co-production between Indigenous knowledge holders and scientists has emerged from several decades of earlier work within ethnoscience and on Indigenous knowledge of the environment. In recent years, however, Indigenous knowledge has become a victim of its own success. The multiplication of calls to include Indigenous knowledge in programmes, projects and research has led to the widespread adoption of the term, but often with little effort to ensure meaningful content. Out of concern that this trend might undermine the legitimacy of Indigenous knowledge, we have formulated the more ambitious project of DKC that is presented here.

2 *Co-production: Founding Concepts and Ethical Frameworks*

Elinor Ostrom, a political economist, was the first to coin the term 'co-production', which was subsequently taken up in science and technology studies. We discuss how these earlier usages inform our own, while also recognizing important insights from reflexive anthropology, radical feminism and research decolonization.

3 *A Methodology and Ethic for Decolonized Knowledge Co-production (DKC)*

We outline the conditions under which this ambitious undertaking might meet with success today, including the need for a problem-oriented approach with focused outcomes; long-term engagement and intellectual rigour to overcome differences in ontologies and epistemologies; and a decolonization of the quest for knowledge so as to address asymmetries of power and achieve equity.

4 *About This Volume*

We conclude with a brief overview of the organization of this book and the papers presented in the following chapters.

Origins: Ethnoscience and Indigenous Knowledge of the Environment

The concept of Indigenous-scientific knowledge co-production builds upon several decades of work on Indigenous, traditional or local knowledge of the environment. For some researchers, the initial approach to Indigenous knowledge has been through the domain of ethnoscience (Conklin, 1954; Sturtevant, 1964). For others, work on 'traditional ecological knowledge' (TEK) emerged in the political context of Indigenous claims to land and resource rights (Freeman, 1976). Despite their distinct origins, these paths have interconnected in subsequent decades and have been mutually reinforcing, creating conditions for the emergence of knowledge co-production.

Ethnoscience: Indigenous Knowledge of 'Natural' Environments

Ethnoscience or the 'new ethnography' – ethnobotany, ethnozoology and ethnoecology – emerged and prospered as a field of research from the 1950s to the early 2000s. Harold Conklin (1954) was a trailblazer, documenting the vast botanical knowledge of the Hanunoo, swidden farmers of the Philippines. In a completely different milieu and culture, Richard Nelson (1969) also provided an early and decisive contribution on Alaskan Iñupiat (Eskimo) knowledge of the physical environment, particularly that of Arctic sea ice. Their pioneering research unveiled the meticulous quality, remarkable amplitude and systematic nature of Indigenous knowledge of the natural milieu, including its transmission from one generation to the next and its parallels and contrasts with Western science. In the 1960s, Levi-Strauss (1962) offered a first overview of research into Indigenous knowledge systems in his seminal book, *The Savage Mind*.

Preparing the way for the cognitive sciences, ethnoscientists focused research on Indigenous systems of denomination and classification of natural phenomena – plants, animals, colours, celestial bodies, etc. – setting in place an ambitious interdisciplinarity involving ethnology, linguistics and natural sciences (Sturtevant, 1964; Berlin et al., 1966; Ellen, 1983, 2006; Toledo, 1992; Friedberg, 1999). Indeed, researchers strove to juxtapose the in-depth knowledge of each culture – the syntax and semantics of Indigenous terms, the signifiers and the signified, social relations, practices and ontologies – with scientific understandings of the plants, animals or other elements under scrutiny. Researchers from many disciplines, including anthropologists, ecologists and geographers, conducted research under various labels: ethnobotany, ethnozoology, ethnobiology, ethnoecology, and more recently, historical ecology (Balée, 2002) and ethnoclimatology (Orlove et al., 2002).

Indigenous Land Claims and the Emergence of TEK and Co-management

Ethnoscience refuted notions prevailing from colonial times about the 'cognitive limitations' of 'primitive' peoples and swept aside colonial ethnobotany with its narrow focus on economically useful plants. However, its focus was on similarities and differences in human cognition rather than on human rights. In contrast, the driving force behind the recognition of Indigenous knowledge of the environment in northern North America was the politics of Indigenous peoples' rights to land and resources. Starting in the late 1960s in Alaska, United States and the 1970s in Canada, legal recognition of Indigenous rights over traditional territories that had not yet been extinguished by earlier treaties (Burger, 1977) opened an era of negotiations between governments and Indigenous peoples. These negotiations addressed not only the delimitation of Indigenous lands (Freeman, 1976) but also aboriginal rights to the harvesting and management of natural resources (Berkes, 1982, 1999). It is in this context that research on traditional ecological knowledge (TEK) emerged, attesting to the extensive and detailed environmental knowledge underlying Indigenous resource use (Freeman and Carbyn, 1988; Berkes, 1999; Nakashima and Roué, 2002; Huntington, 2011) and Indigenous self-management regimes (Feit, 1973).

These land claim processes led to the establishment of wildlife co-management regimes with Indigenous peoples seated alongside government representatives to share knowledge, priorities and practices (Pinkerton, 1989; JBNQA, 1998). The term 'co-production of knowledge' made its first appearance as part of an effort to include the knowledge of the Dene peoples and biologists in modelling caribou availability in north-western Canada and Alaska.

We strive towards the co-production of knowledge with communities … not to meld cultural perspectives, but in an attempt to improve communication among parties and resolve common problems.

Kofinas and Braund, 1998: 3

This association of co-management with knowledge co-production is made explicit in Armitage et al. (2011), who demonstrate how the institutional arrangements of co-management may be conducive to the successful co-production of knowledge (Nakashima et al., 2012). The authors interpret co-production as a social learning process among a wide circle of organizations and individuals, both Indigenous and government. The close ties between co-production and co-management are also the subject of a recent special issue of the journal *Arctic Science*, the editors of which emphasize that 'the meaningful inclusion of Inuit and their knowledge systems is both implicit and explicit in wildlife co-management' (Johnson et al., 2020: 124).

Arctic and subarctic North America have been at the forefront of these developments as a result of political pressures to settle Indigenous land claims in the face of oil and gas, hydropower and mineral exploitation. However, recognition of Indigenous knowledge and resource management is not by any means restricted to these regions. Research among peoples of the Pacific Islands reveals their exceptional knowledge of the marine environment and the wealth of traditional institutions that manage resource access and conserve marine ecosystems in finite island environments (Johannes, 1978, 1981; Ruddle and Johannes, 1984). Similar ground-breaking work has been conducted in Africa (Scoones and Thompson, 1994; Warren et al., 1995), Asia and Latin America (Sillitoe, 2007).

Indigenous Knowledge: Growing Recognition and the Quagmire of Political Correctness

This growing local and national recognition of the value of knowledge from Indigenous cultures and environments across the globe has contributed to its emergence at the international level as an essential component of humanity's response to planetary environmental crises. The United Nations Convention on Biological Diversity (CBD) provided a decisive push in 1992 with its recognition of the importance of 'traditional knowledge, innovations, and practices' under its Article 8(j). In subsequent years, Indigenous knowledge came to be an established feature of an ever-expanding web of research programmes, development initiatives and institutional arrangements. Prominent recognition was provided by scientific expert bodies such as the Intergovernmental Panel on Climate Change (IPCC) (beginning with its Third Assessment Report in 2001) and the Intergovernmental Platform for Biodiversity and Ecosystem Services (IPBES) since its initial session in 2013 (Nakashima et al., 2018).

This expansion of Indigenous knowledge into a multitude of arenas at national and international levels held much promise for Indigenous peoples. It opened the possibility that their knowledge, insights and priorities would be considered, alongside science, in a wide range of environmental decision-making processes. However, it also triggered a proliferation of superficial and opportunistic allusions to the concept.

Numerous programmes and projects have tacked on Indigenous knowledge components in order to tick the box of stakeholder involvement that many donors require. Nevertheless, without the necessary expertise or sufficient motivation to genuinely engage with local communities, these responses steeped in political correctness do little good and considerable harm. Instrumentalized in this manner, Indigenous knowledge is sapped of its meaning and, in some instances, the term has become little more than a buzzword.

This widespread recognition has nonetheless enabled the emergence of a new phase of knowledge collaboration. Indigenous peoples and scientists are joining forces in decolonized partnerships to grapple with the complex challenges posed by global change. We designate these emerging collaborations across knowledge systems as *decolonized knowledge co-production between Indigenous knowledge holders and scientists (DKC)*.

Co-production: Founding Concepts and Ethical Frameworks

Earlier applications of the term are briefly presented here to better understand how they inform our concept of co-production between Indigenous knowledge and science.

Ostrom's Co-production of Public Services: From Observation to Aspiration

Elinor and Vincent Ostrom, a political economist and a political scientist, respectively, coined the term 'co-production' in the context of their work on polycentric public services. Examining the delivery of security by local and centralized police departments in and around the city of Indianapolis, United States (Ostrom and Whitaker, 1973), Ostrom et al. (1978) concluded that citizens should be recognized as 'co-producers with police' of public security as they play an essential role in the delivery of outcomes including the success of investigations, number of arrests and rates of criminality (Ostrom et al., 1978: 383). The authors themselves point out that co-production is 'a rather novel and important aspect of our approach' (Ostrom et al., 1978: 389). Indeed, the concept features prominently in subsequent work with additional examples such as the co-production of education by teachers and students, and of health by doctors and patients. In a paper that spells out the methods of institutional analysis, Ostrom (1985) describes the conditions leading to co-production,

... when the outcomes of a process cannot be produced without the active cooperation of several different owners of input resources, the outcomes are subject to co-production.

Ostrom, 1985: 13

Whereas co-production is initially presented as a social phenomenon deduced from field observations, it subsequently emerges as a central tenet of Ostrom's vision of 'the Good Society' (Ostrom, 1993). She rails against the increasingly pervasive notion of one-way, top-down governance where governments are responsible for service delivery and citizens are confined to the passive role of consumers or clients.

The term "client" is used more and more frequently to refer to those who should be viewed as the essential co-producers of their own education, safety, health and communities. A client is the name for a passive role. Being a co-producer makes one an active partner.

Ostrom, 1993: 8

For Ostrom, co-production is an essential building block for democratic societies. It is an integral part of two-way governance processes involving citizens who, as co-producers, 'take responsibility for as much as possible of what happens around them' (Ostrom, 1993: 8). She expressed her fear that this active role is increasingly threatened by the growing hegemony of centralized government.

Rabeharisoa and Callon (2004) describe one such co-production process where patients and their families engage medical scientists in a mutual apprenticeship (Epstein, 1995, 1998) to advance medical knowledge about rare and orphan diseases. This 'new relationship between science and society, between those who produce knowledge and those who are supposed to benefit' from it (Rabeharisoa and Callon, 2002: 71) is driven by patient associations such as the French Association of Myopathies that defend the right to 'self-help' and lobby for taking into account the experiences of the patients themselves. Embracing this concept of knowledge co-production, one such collective of patients, doctors, S&T experts and artists have named themselves 'The Institute of Knowledge Co-production on the Huntington Disease Dingdingdong' (Rivière, 2013).

Co-production of Science and Society: Insights from Science and Technology Studies

Co-production, as conceived by Ostrom, focuses on the delivery of public services by governments and citizens. Philosophers and epistemologists in science and technology studies (STS), however, have considered the particular case of science: its co-production as an integral part of society and its assertions about 'truth' and 'reality'.

Physician and biologist Ludwick Fleck is the first to have argued that scientific concepts and theories are culturally conditioned. His iconoclastic book 'Genesis and Development of a Scientific Fact' published in 1935, continues to fluster scientists and their claims to 'truth' (Fleck, 1979). He demonstrates that 'facts' are constructed by groups of scientists who, as 'thought collectives' with specific 'thought-styles', create and adhere to norms, conceptions and practices that differ from those of other scientists in the same discipline. Scientific 'facts' produced by one thought collective are incommensurable with those of another, thus unveiling the relative nature of 'truth'.

Thomas Kuhn (1962) takes issue with the positivist epistemology of Karl Popper, for whom science is cumulative and progresses in linear fashion. He postulates, instead, that science is cyclic. Following a period of 'normal (cumulative) science', a crisis is encountered that engenders a 'gestaltswitch' or *paradigm shift*, after which a new period of 'normal' science is set in motion until the next paradigmatic crisis. This change in paradigms is so profound that the science existing before the crisis is, to a large extent, incommensurable with the science that follows. It is a shift from one way of viewing the world to another.

Fleck reveals the extent to which scientific 'facts' are socially constructed by specific groups of scientists, while Kuhn demonstrates that scientific 'truths' shift historically from one period to the next. This 'social construction of scientific facts' is further analysed by Bruno Latour (1990), who for the first time outlined a 'theory of the co-production of science and its social context' in his review of 'Leviathan and the Air Pump', Shapin and Schaller's book on the seventeenth-century debate about the air pump. For Latour, science is but one of several 'modes of existence' that co-exist among the 'moderns' of contemporary Western society, each incommensurable with the other because each 'possess(es) its own conditions of truth and falsity' (Latour, 2013: 177). In view of this incommensurability, Latour postulates the need for a system of 'diplomacy' that allows proponents to interact without being judged by the regime of 'veridiction' (truth and falsity) of the other (Latour, 2013).

Sheila Jasanoff reflects further on the term co-production in an essay on the sociology of scientific knowledge that reaffirms that science is 'a dynamic and integral part of society – a social construct' (Jasanoff 1996: 409).

Scientific knowledge, in particular, is not a transcendent mirror of reality. It both embeds and is embedded in social practices, identities, norms, conventions, discourses, instruments and institutions – in short, in all the building blocks of what we term the social.

Jasanoff, 2004: 2–3

By re-situating science within society, rather than as an imagined extra-societal undertaking, Jasanoff debunks the binary oppositions that Western science has long used to extol its own virtues as objective and rational, while disparaging other knowledge systems as value-laden and rooted in superstition.

Co-production can therefore be seen as a critique of the realist ideology that permanently separates the domains of nature, facts, objectivity, rationality and policy from those of culture, value, subjectivity, emotion and politics.

Jasanoff, 2004: 2–3

However, rather than acknowledging that science is co-produced with society, it is much more tempting for scientists to perpetuate these culturally constructed oppositions of nature versus culture or facts versus values. For Isabelle Stengers, the renowned science philosopher, science professors are complicit even today in promulgating this cognitive hierarchy among their students.

I have learned from my teaching experience that most students enrolled in the so-called 'hard' sciences are determined to forget their courses (in epistemology of science) once exams are over. ... Although obliged to put up with these courses that they consider 'chatty', they do not view them as a crucial part of their curriculum, an attitude seconded by many of their 'true' professors with a smirk, complicit smile or wise advice about not allowing oneself to 'be dispersed'.

Stengers, 2013: 16–17 (author's translation)

Recognizing that science is inextricably linked to society is a crucial first step towards its decolonization, and is essential for achieving, as we will discuss subsequently, a decolonized co-production of science and Indigenous knowledge.

From Postcolonialism to Decolonization

Other writings lay foundations for the emergence of knowledge co-production by revolting against the scientist's reification of their 'research objects'. Indigenous intellectuals have triggered a decolonization of research ethics, initially among the social sciences and, more recently, among parts of the biological sciences. At the time the American Indian Movement was coming into being, Vine Deloria, an Indigenous intellectual, accused anthropologists of reducing Indigenous peoples to being mere 'objects for observation' (Deloria, 1997). In a book entitled 'Custer Died for Your Sins: An Indian Manifesto', which gained considerable renown, he asks: 'Why should we continue to be the private zoos for anthropologists?' (Deloria 1970: 99). As aptly summarized by Biolsi and Zimmerman:

Deloria represented the anthropologist as an urban, overly intellectualized, insufficiently humanized academic who descends on Indian country every summer to confirm and reproduce essentially self-confirming, self-referential and self-reproducing closed systems of arcane "pure knowledge" – systems with little, if any, empirical relationship to, or practical value for, real Indian people.

Biolsi and Zimmerman, 1997: 2

Similarly, Edward Saïd (1978), in his book *Orientalism*, denounces Western discourses about the Middle East as justifications for their colonial enterprises. By demonstrating how the 'Other' is invented by the Western world in order to fulfil their own needs, he becomes the leader of a new movement: post-colonialism.

In the decade that followed, reflexive anthropology, semiotic anthropology and post-structuralism dedicated themselves to redefining objectivity in relation to knowledge and the subject/object relationship. Feminists in particular criticize the reifying stance of science, proposing in its place a situated and embodied knowledge (Haraway, 1988: 583). They advocate a 'positioned rationality' and lambast scientists who, by presenting themselves as the holders of a universal objectivity, place themselves in a position of cultural hegemony, forgetting to analyse their own worldviews: 'There is no way to be simultaneously in all, or wholly in any, of the privileged (i.e. subjugated) positions structured by gender, race, nation, and class' (Haraway, 1988).

Reflexive anthropology, founded by Rabinow (1983) and Clifford and Marcus (1986), abandons the position of scientists as subjects who study objects, proposing in its stead a dialogue inspired by the notion of dialogism invented by Bakhtin in the domain of literature,

Reflexive anthropology sees the resultant production as a dialogue between anthropologist and informant so-called: the observer/observed relationship can no longer be assimilated to that between subject and object. The object(ive) is a joint production. Many voices, multiple texts, plural authorship.

Strathern, 1987: 264, note 38

Following in the footsteps of Edward Saïd, Linda Tuhiwai Smith published *Decolonizing Methodologies: Research and Indigenous Peoples* in 1999. She argues that collaborations with non-Indigenous scientists are difficult to envisage as 'there is no difference (. . .) between "real" or scientific research and other visits by inquisitive and acquisitive strangers' (Tuhiwai Smith, 1999: 3). Her critique, which targeted extractive research and an anthropology that reified the people studied, has created an opportunity to establish a new relationship.

Whose research is it? Who owns it? (. . .) Who will benefit from it? Who has designed its questions and framed its scope?

Tuhuwai Smith, 1999:10

The questions she poses and the requirements she defines could be understood as a programmatic forerunner to what we propose as a decolonized co-production between Indigenous and scientific knowledge systems.

A Methodology and Ethic for Decolonized Knowledge Co-production between Indigenous Knowledge Holders and Scientists (DKC)

As we have seen in the previous section, the term co-production is polysemic. In the first instance, it is *descriptive* as it describes a process inherent to a given society. It might also be termed *intrinsic co-production* in reference to it being an essential or fundamental component of, for example, public services (Ostrom et al., 1978) or science (Jasanoff, 1996). But the term may also be *prescriptive*, when co-production is put forward as an aspiration or norm, such as for democratic societies (Ostrom, 1993) or to advance knowledge about rare or emerging diseases (Rabeharisoa and Callon, 2002). Furthermore, Meadow et al. (2015) underline the need for a *deliberate co-production* between scientists and decision-makers that is explicitly planned and executed in an iterative and reflexive manner. Our proposal goes one step further in that it concerns a deliberate co-production, not between groups within the same society, but between scientists from diverse disciplines and traditional knowledge experts from Indigenous societies.

This cross-cultural co-production is recent in origin as its emergence is predicated on scientists engaging in a reflexive analysis of Western science and their own scientific practice, and a progression from descriptive co-production to one that is prescriptive and then deliberate. Therefore, rather than addressing knowledge co-production 'in general', this book focuses on co-production between two specific groups, Indigenous knowledge holders and scientists, in response to environmental risks and uncertainties unleashed by global change, including anthropogenic perturbation of climate and life systems. At the advent of the Anthropocene, these are as yet uncharted waters. Neither science nor Indigenous knowledge can provide answers for challenges that are only beginning to emerge. Yet these knowledge systems are largely complementary: science focuses on large-scale, even global phenomena and trends, with an appetite for quantification and the universal; while Indigenous systems are rooted in intimate, qualitative understandings of the inner workings of local systems. In the face of uncertainty, observations, understandings and interpretations from both systems may be usefully shared and jointly debated to co-produce new knowledge that may lead to effective responses.

Articulating two knowledge systems, however, is not just a matter of inventing a new method and a strategy for its global application. Knowledge is intimately connected with power (Pohl et al., 2010), and in our contemporary societies, a severe power asymmetry persists between science and Indigenous knowledge. Even though Indigenous peoples across the globe have hunted, fished, gathered, herded or farmed their homelands on land and sea for millennia, their environmental knowledge has long been ignored, if not disdained.

State-management systems with science-based policies of quotas, closed seasons, protected areas and maximum yields have profoundly transformed that which we call 'nature' in Western cultures. The opposition of nature with culture is a unique trait of our naturalist ontology (Descola, 2005). Such a dualism does not exist for Indigenous peoples for whom human and non-human persons are part of the same continuum. The imposition of Western rules and regulations on Indigenous homelands (often misrepresented as 'nature' or 'wilderness') threatens their cultures, ways of life, and their very existence. Without addressing these asymmetries of power and differences in ontologies, efforts to build dialogue between Indigenous knowledge and science run the risk of merely reinforcing this deeply rooted colonial heritage.

DKC is ambitious, but it is not by any means a *merging* of *entire* knowledge systems. Knowledge systems are rooted in unique cosmologies and epistemologies, as well as their own social dynamics and ways of life. They cannot be blended whole scale. The goal that we envisage for knowledge co-production is more modest. Its aim is not to merge, but rather to seek out correspondences, dissonances and complementarities between two distinct but intersecting views on jointly defined and clearly circumscribed problems. Such a process, freed of the inequities instilled by our colonial past, may lead to a collaborative generation of new knowledge and novel approaches to help resolve the complex socio-ecological challenges facing humanity today.

Although encouraging advances have been made during recent decades, decolonized knowledge co-production between Indigenous knowledge and science remains an ambitious undertaking that few, if any, can claim to have fully achieved. Some of the essential requirements to progress towards this goal are as follows:

Prerequisites

A Long-Standing Dialogue

DKC can only emerge from a long-term exchange between Indigenous and scientific experts that is measured in years, if not in decades. In sharp contrast, Rapid Rural Appraisal (RRA) and other participatory processes use rapidity as the measure of their efficacy and the inevitable consequence is superficiality (Richards, 1995).

An In Situ *Approach Anchored in Local Knowledge of Indigenous Homelands*

DKC occurs *in situ*, within the home territories of the knowledgeable Indigenous experts who partner with field scientists. It is built around their in-depth local knowledge and relevant sciences.

A Problem-Oriented and Engaged Approach

DKC is a circumscribed undertaking that focuses on a mutually agreed problem faced by Indigenous communities. By focusing dialogue and debate on matters essential to the defined problem, conditions conducive to successful co-production between Indigenous knowledge and science are created.

An Intentional and Applied Undertaking with Theoretical Implications

By fostering innovative methodologies and the generation of new knowledge, DKC is an applied undertaking that may also have important theoretical implications, for both Indigenous knowledge holders and scientists.

Broad Interdisciplinarity and Transdisciplinarity

DKC requires much more than a multidisciplinary approach, which merely juxtaposes scientific disciplines without resolving the fundamental problem posed by their different temporal and spatial scales. To consider complex issues such as the local impacts of global phenomena, a broad interdisciplinary approach is needed that not only engages different disciplines in a shared undertaking in order to address a jointly defined problem but also bridges the domains of the bio-physical and social sciences. Even more challenging is the establishment of dialogue that extends beyond science to encompass both scientific and Indigenous knowledge, often referred to as transdisciplinarity.

Key Actors

A Community of Practice

In DKC, each project requires the constitution of a community of practice that works to bridge the divide between knowledge systems and communities with their distinct cultures, languages, methods and worldviews. Strict respect for equity between knowledges and an absence of hierarchy in social interactions are essential requirements for an effective dialogue that remains attentive to ontological and semantic conflicts which may offer critical insights. Finally, personal ties established over time as the work progresses create a climate of mutual confidence that ensures the success of the endeavour.

Indigenous Experts

DKC requires the involvement of individuals recognized by their communities as experts in the targeted domain, whether their recognized expertise is thematic, spatial (knowledge about particular places), or temporal (generations with knowledge from different periods). Both expert women and men are involved for their gender-based knowledge.

Scientific Experts

In DKC, to establish and maintain productive dialogue with Indigenous experts, scientists must be knowledgeable about the bio-physical issue at hand, whether it might relate to climate conditions, ice dynamics or ecological change. Social science expertise in Indigenous languages and ethnoscientific methods is also fundamental in order to decipher semantic conflicts and steer clear of ontological and epistemological pitfalls, along with transdisciplinary expertise between science and other knowledge systems. The most important requirements, however, are of a completely different order: reflexivity, humility and a passion for knowledge that renounces the hegemony of science are essential prerequisites for establishing a constructive dialogue across cultural boundaries.

Mediators

DKC requires capable and experienced mediators, both Indigenous and scientific, to bridge the divide between knowledge systems. Their key role is to serve as translators, not in the linguistic sense, but rather in the ontological sense of finding common ground in relation to meanings. This critical function generally requires long experience of collaborative interactions with persons from both knowledge systems.

Ethics and Relationships

Decolonization

DKC at the interface of scientific and Indigenous knowledge systems requires new partnerships rooted in an epistemic decolonization that is gaining momentum. Indigenous rights are formally recognized by the 2007 UN Declaration on the Rights of Indigenous Peoples. Indigenous knowledge holders have gained broad international recognition, building on the platform provided by the Convention on Biological Diversity and its Article 8(j), further reinforced by the Nagoya Protocol on Access and Benefit-Sharing. A new generation of Indigenous leaders has emerged that is at home in two worlds, combining their training in Western institutions with knowledge from their communities. Among interdisciplinary scientists specializing in Indigenous knowledge, some have moved beyond merely documenting and valuing this intellectual heritage to partnering with and for Indigenous peoples.

Equity between Knowledge Systems

DKC requires the dismantling of hierarchical relations between science and Indigenous knowledge systems. While in citizen science, scientists alone fix research questions, establish methods, analyse data and draw conclusions, co-production requires a level playing field and equitable partnership from day one. There is no place for the 'validation' of one knowledge system by the other (Roué and Nakashima, 2018) or for an extractivist approach where one system selects knowledge of value and rejects the rest.

Mutual Trust and Benefits

In DKC, trust between Indigenous and scientific partners is born from long experience of joint work that builds the confidence of all parties in a shared capacity to willingly and successfully navigate between distinct languages, classification systems and ontologies: basic requirements for an effective and respectful dialogue. Equally important is the identification of mutual benefits that are clearly defined and agreed from the beginning with shared responsibility for their fullest realization.

From FPIC to Co-authorship

Free Prior and Informed Consent (FPIC) is a right included as Article 10 of the UN Declaration on the Rights of Indigenous Peoples (UNDRIP) of 2007. Not only must Indigenous peoples be consulted prior to any project, but they are also free to withdraw their consent at any point in time. DKC, however, goes well beyond FPIC for project initiation. Indigenous experts are fully engaged and responsible partners alongside scientists throughout the process from the project's earliest stages to its completion. This shared responsibility extends to project outputs, including transparent co-authorship of reports and publications by the Indigenous experts and scientists jointly involved in this equitable enterprise.

About This Volume

This book responds to the need to move beyond reaffirming and documenting the importance of Indigenous peoples' knowledge. It proposes decolonized knowledge co-production between Indigenous knowledge holders and scientists (DKC) as a goal to work towards and presents case studies that illustrate some of the different efforts underway. It is organized in three major parts:

Part I   From Practice to Principles: Methods and Challenges for DKC
Part II   Indigenous Perspectives on Environmental Change
Part III Global Change and Indigenous Responses

Part I From Practice to Principles: Methods and Challenges for Decolonized Knowledge Co-production (DKC)

This first part focuses on research and projects in the Arctic and Subarctic regions that have been on the frontlines of climate change due to the accelerated rate of circumpolar warming (Krupnik and Jolly, 2002). Political processes relating to Indigenous land and resource rights have stimulated several decades of joint work by circumpolar Indigenous peoples and researchers on Indigenous knowledge systems and regimes of co-management. With legal recognition of Indigenous knowledge in northern Canada and Alaska, and ambitious international collaborations such as during the International Polar Year 2007–2008 (Krupnik et al., 2011), long-term partnerships between Indigenous peoples and scientists have been forged over recent decades that offer a unique opportunity for reflexive analyses of DKC (Roturier et al., Chapter 7).

Authors emphasize the importance of a step-by-step and feedback-driven process – learning by doing. The opportunity for co-production emerged as the project advanced, even though it may not have been envisaged from the beginning: 'As trust was built and relationships were created among the participants, people began to realize that more was possible,' (Huntington et al., Chapter 2). Similarly, a broad interdisciplinarity, extending from the physical to the ecological sciences (Druckenmiller, Chapter 4), and also embracing history and the social sciences (Eicken et al., Chapter 3; Krupnik, Chapter 5), is not achieved a priori but rather assembled step-by-step as required by the complexity of the problem to be resolved.

In DKC, long-term relationships based on collaborative 'learning by doing' create fertile ground for serendipity: discovery emanating from the unexpected. This phenomenon, associated with many scientific 'inventions', is often mistakenly represented as 'chance'. But as Louis Pasteur so famously declared, 'Chance only favours prepared minds'. In dialogues between knowledge systems, semantic 'anomalies' emerge where concepts once believed to be synonymous are in fact discovered to differ (Roué et al., Chapter 6).

Part II Indigenous Perspectives on Environmental Change

Part II highlights Indigenous voices. The texts presented here are based on interviews conducted by M. Roué during UNFCCC COP 21 and 22, or as part of the BRISK project. Their shared message in very different settings is about monitoring change and responding with resilience. Conscious that loss of knowledge leads to a loss of identity, they strive to pass on to younger generations traditional knowledge and ways of life as a foundation for adapting to an era of global change. In contrast to popular images of Indigenous peoples resisting modernity, they are open to technical innovations, adapting them to fit their needs and maintain their modes of existence. They continually reinvent tradition to keep it alive, whether for wayfaring in the Pacific (Fuluna, Chapter 9), herding reindeer in Sapmi (Bongo, Chapter 10) or raising livestock in the Sahel (Dicko, Chapter 11).

For those who believe that Indigenous rights are upheld at the international level, the account by Oumarou (Chapter 8) illustrates how difficult it is for Indigenous voices to be heard in UN arenas where only governments have the right to speak. Capable of defending Indigenous rights in the international arena with enormous determination and great talent, while also working on the ground with members of her herding community, the experience of this young Indigenous woman is exemplary. It resonates with that of Indigenous peoples around the globe who rise to challenges that are at once global and local.

Part III Global Change and Indigenous Responses

The third part of the book illustrates the enormity of the challenges facing knowledge co-production in a world where climate change impacts are further aggravated by global change. While some Indigenous efforts to adapt may meet with success (Steward, Chapter 14; Fernandes-Llamares and Cabeza, Chapter 15), skewed power relations may complicate final outcomes. Their efforts may end up appropriated and monetized by government, as in the Himalayas (Salick, Chapter 12) or compromised by industrial development, as in the case of Nenets reindeer herders who co-exist with the largest natural gas complex in the Arctic (Forbes, Chapter 13). These cases remind us of how illusory it is to believe that simple technical fixes would suffice to counter climate risks, when in fact Indigenous peoples worldwide are facing complex challenges of global change, failed governance and new forms of coloniality (Burman, Chapter 16). In this rapidly changing world, a decolonized co-production that unites the forces of Indigenous knowledge with that of science may provide Indigenous peoples with much-needed additional leverage.

Acknowledgements

Sincere thanks to Igor Krupnik for his valuable comments that helped us shape this chapter. We are also grateful for the financial support provided by the Agence Nationale de la Recherche (ANR-France) for the programme Bridging Indigenous and Scientific Knowledge (BRISK) and the programme Future Arctic Ecosystems (FATE) in the framework of the Belmont Forum.

References

Adelle, C., Pereira, L., Gorgens, T. and Losch, B. 2020. Making sense together: The role of scientists in the coproduction of knowledge for policy making. *Science and Public Policy*, 47 (1): 56–66. https://doi.org/10.1093/scipol/scz046

Armitage, D., Berkes, F., Dale, A., Kocho-Schellenberg, E. and Patton, E. 2011. Co-management and the co-production of knowledge: Learning to adapt in Canada's Arctic. *Global Environmental Change*, 21(3): 995–1004. https://doi.org/10.1016/j.gloenvcha.2011.04.006

Balée, W. L. (ed.). 2002. *Advances in Historical Ecology*. New York, Columbia University Press.

Berkes, F. 1982. Waterfowl management and northern Native Peoples with reference to Cree Hunters of James Bay. *Musk-Ox : A Journal of the North*, 30: 23–35.

Berkes, F. 1999. *Sacred Ecology: Traditional Ecological Knowledge and Resource Management*. Philadelphia, Taylor and Francis.

Berlin, B., Breedlove, D. E. and Raven, P. H. 1966. Folk taxonomies and biological classification. *Science*, 154: 273–275.

Biolsi, T. and Zimmerman, L. J. 1997. Introduction: What's changed and what hasn't. pp. 2–23 In Biolsi, T. and Zimmerman, L. J. (eds.). *Indians and Anthropologists: Vine Deloria, Jr. and the Critique of Anthropology*. Tucson, University of Arizona Press.

Burger, T. R. 1977. *Northern Frontier, Northern Homeland: The Report of the Mackenzie Valley Pipeline Inquiry*. Toronto, James Lorier and Co.

Clifford, J. and Marcus, G. (eds.). 1986. *Writing Culture: The Poetics and Politics of Ethnography*. Berkeley, University of California Press.

Conklin, H. 1954. *The Relation of Hanunoo Culture to the Plant World*. New Haven, CT, Yale University Press.

Deloria, V. 1970 [1969]. *Custer Died for Your Sins: An Indian Manifesto*. New York, Avon.

Deloria, V. 1997. *Red Earth, White Lies: Native Americans and the Myth of Scientific Fact*. Golden, CO, Fulcrum Publishing.

Descola, P. 2005. *Par-delà Nature et Culture*. Paris, Editions Gallimard.

Ellen, R. 1983. *The Cultural Relations of Classification: An Analysis of Nuaulu Animal Categories from Central Seram*. Cambridge, Cambridge University Press.

Ellen, R. 2006. *The Categorical Impulse: Essays in the Anthropology of Classifying Behavior*. New York, Berghahn.

Epstein, S. 1995. The construction of lay expertise: AIDS activism and the forging of credibility in the reform of clinical trials. *Science, Technology & Human Values*, 20(4): 408–437. https://doi.org/10.1177/016224399502000402

Epstein, S. 1998. *Impure Science: AIDS, Activism and the Politics of Knowledge*. Berkeley, University of California Press.

Feit, H. 1973. Ethno-ecology of the Waswanipi Cree: Or how hunters can manage their resources. pp. 115–125. In Cox, B. (ed.). *Cultural Ecology*. Toronto, McClelland and Steward.

Fleck, L. 1979 [1935]. *Genesis and Development of a Scientific Fact*. Translated by Bradley F. and Trenn T. J. Chicago, University of Chicago Press.

Freeman, M. M. R. (ed.). 1976. *Inuit Land Use and Occupancy Project*. Vols. I to III. Ottawa, Minister of Supplies and Services Canada.

Freeman, M. M. R. and Carbyn L. N. (eds.). 1988. *Traditional Knowledge and Renewable Resource Management in Northern Regions*. Edmonton, Boreal Institute for Northern Studies.

Friedberg, C. 1999. Diversity, order, unity: Different levels in folk knowledge about the living. *Social Anthropology*, 7: 1–16. https://doi.org/10.1017/S0964028299000014

Haraway, D. 1988. Situated knowledges: The science question in feminism and the privilege of partial perspective. *Feminist Studies*, 14(3): 575–599. https://doi.org/10.2307/3178066

Huntington, H. 2011. The local perspective. *Nature*, 478: 182–183. https://doi.org/10.1038/478182a

Jasanoff, S. 1996. Beyond epistemology: Relativism and engagement in the politics of science. *Social Studies of Science*, 26: 393–418. https://doi.org/10.1177/030631296026002008

Jasanoff, S. (ed.) 2004. *States of Knowledge: The Co-Production of Science and Social Order*. London, Routledge.

JBNQA. 1998. *James Bay and Northern Quebec Agreement and Complementary Agreements*. Quebec, Les Publications du Quebec.

Johannes, R. E. 1978. Traditional marine conservation methods in Oceania and their demise. *Annual Review of Ecology and Systematics*, 9: 349–364. https://doi.org/10.1146/annurev.es.09.110178.002025

Johannes, R. E. 1981. *Words of the Lagoon: Fishing and Marine Lore in the Palau District of Micronesia*. Berkeley, CA, University of California Press.

Johnson, N., Pearce, T., Breton-Honeyman, K., Etiendem, D. N. and Loseto, L. L. 2020. Knowledge co-production and co-management of Arctic wildlife: Editorial. pp. 124–126. In Knowledge mobilization on co-management, co-production of knowledge, and community-based monitoring to support effective wildlife resource decision making and Inuit self-determination. Special Issue. *Arctic Science*, 6: 124–360.

Kofinas, G. and Braund, S. 1998. *Local Caribou Availability: A Draft Report from Community Involvement Phase 2*. Institute of Social and Economic Research, University of Alaska Anchorage.

Krupnik, I. and Jolly, D. (eds.). 2002. *The Earth Is Faster Now: Indigenous Observations of Arctic Environmental Change*. Fairbanks, AK: ARCUS.

Krupnik, I., Allison, I., Bell, R., Cutler, P., Hik, D., López-Martínez, J., Rachold, V., Sarukhanian, E. and Summerhayes, C. (eds.). 2011. Understanding Earth's Polar Challenges: International Polar Year 2007–2008. University of the Arctic: Rovaniemi, Finland, CCI Press: Edmonton, Alberta, Canada and ICSU/WMO Joint Committee for International Polar Year 2007–2008.

Kuhn, T. S. 1962. *The Structure of Scientific Revolutions*. Chicago, University of Chicago Press.

Latour, B. 1990. Postmodern? No simply amodern. Steps towards an anthropology of science: An essay review. *Studies in the History and Philosophy of Science*, 21: 145–171.

Latour, B. 2013. *An Inquiry into Modes of Existence: An Anthropology of the Moderns*. Cathy Porter (tr.). Cambridge, MA, Harvard University Press.

Lemos, M. C., Arnott, J. C., Ardoin, N. M. et al. 2018. To co-produce or not to co-produce? *Nature Sustainability*, 1: 722–724. https://doi.org/10.1038/s41893-018-0191-0

Levi-Strauss, C. 1962. *La pensée sauvage* (The Savage Mind). Paris, Plon.

Meadow, A. M., Ferguson, D. B., Guido, Z., Horangic, A., Owen, G. and Wall. T. 2015. Moving toward the deliberate coproduction of climate science knowledge. *Weather, Climate, and Society*, 7: 179–191. https://doi.org/10.1175/WCAS-D-14-00050.1

Miller, C. A. and Wyborn, C. 2018. Co-production in global sustainability: Histories and theories. *Environmental Science and Policies*. https://doi.org/10.1016/j.envsci.2018.01.016

Nakashima, D. and Roué, M. 2002. Indigenous knowledge, peoples and sustainable practice. pp. 314–324. In Munn, T. (ed.). *Encyclopaedia of Global Environmental Change*. Chichester, John Wiley & Sons.

Nakashima, D., Krupnik, I., and Rubis, J. T. 2018. *Indigenous Knowledge for Climate Change Assessment and Adaptation*. Cambridge, University of Cambridge Press and Paris, UNESCO Publishing.

Nakashima, D. J., Galloway McLean, K., Thulstrup, H. D., Ramos-Castillo, A. and Rubis, J. T. 2012. *Weathering Uncertainty: Traditional Knowledge and Climate Change Assessment and Adaptation*. Knowledges of Nature 5. Paris, UNESCO and Darwin, UNU.

Nelson, R. K. 1969. *Hunters of the Northern Ice*. Chicago, University of Chicago Press.

Orlove, B., Chiang, S., John, C. H. and Cane, M. A. 2002. Ethnoclimatology in the Andes. *American Scientist*, 90: 428–435. https://doi.org/10.1511/2002.5.428

Ostrom, E. 1985. *Formulating the Elements of Institutional Analysis*. Paper presented at a conference on Institutional Analysis and Development, Washington, DC, Studies in Institutional Analysis and Development.

Ostrom, E. 1993. Covenanting, co-producing and the good society. *PEGS Newsletter*, 3: 7–9.

Ostrom, E. and Whitaker, G. 1973. Does local community control of police make a difference? Some preliminary findings. *American Journal of Political Science*, 17: 48–76. https://doi.org/10.2307/2110474

Ostrom, E., Parks, R. B., Whitaker, G. P. and Percy, S. L. 1978. The public service production process: A framework for analyzing police services. *Journal of Policy Studies*, 7: 381–389.

Pinkerton, E. (ed.). 1989. *Cooperative Management of Local Fisheries: New Directions for Improved Management and Community Development*. Vancouver, University of British Columbia Press.

Pohl, C., Rist, S., Zimmermann, A., Fry, P., Gurung, G. S., Schneider, F., Speranza, C. I., Kiteme, B., Boillat, S., Serrano, E., Hirsch Hadorn, G. and Wiesmann, U. 2010. Researchers' roles in knowledge co-production: Experience from sustainability research in Kenya, Switzerland, Bolivia and Nepal. *Science and Public Policy*, 37(4): 267–281. https://doi.org/10.3152/030234210X496628

Rabeharisoa, V. and Callon, M. 2002. L'engagement des associations de maladies dans la recherche. *Revue Internationale des Sciences sociales*, 171: 65–73. https://doi.org/10.3917/riss.171.0065

Rabeharisoa, V. and Callon, M. 2004. Patients and scientists in French muscular dystrophy research. pp. 142–160. In Jasanoff, S. (ed.). 2004. *States of Knowledge: The Co-Production of Science and Social Order*. London, Routledge.

Rabinow, P. 1983. 'Facts are a word of God': An essay review. In Stocking, G. W. (ed.). *Observers Observed*. (History of Anthropology I.). Madison, University of Wisconsin Press.

Richards, P. 1995. Participatory rural appraisal: A quick and dirty critique. *PLA Notes*, 24: 13–16 (accessed November 27, 2020) http://pubs.iied.org/pdfs/G01591.pdf

Rivière, A. 2013. *Manifeste de Dingdingdong, précédé De La Chorée (1872) de G. Huntington*. Translated by V. Bergerat. Paris, Editions Dingdingdong.

Roué, M. and Nakashima, D. 2018. Indigenous and local knowledge and science: From validation to knowledge co-production. In Callan, H. (ed.). *The International Encyclopaedia of Anthropology*.

Ruddle, K. and Johannes, R. E. (eds.). 1984. *The Traditional Knowledge and Management of Coastal Systems in Asia and the Pacific*. Jakarta, UNESCO.

Saïd, E. W. 1978. *Orientalism*. New York, Pantheon Books.

Scoones, I. and Thompson, J. 1994. *Beyond Farmer First: Rural People's Knowledge, Agricultural Research and Extension Practice*. London, Intermediate Technology.

Sillitoe, P. (ed.) 2007. *Local Science vs. Global Science: Approaches to Indigenous Knowledge in International Development*. New York, Berghahn Books.

Smith, L. T. 1999. *Decolonizing Methodologies: Research and Indigenous Peoples*. Dunedin and London, Zed Books and University of Otago Press.

Stengers, I. 2013. *Une autre science est possible !* Paris, Ed. La Découverte.

Strathern, M. 1987. Out of context: The persuasive fictions of anthropology. *Current Anthropology*, 28(3): 251–281.

Sturtevant, W. 1964. Studies in Ethnoscience. *American Anthropologist*, 66: 99–131. https://doi.org/10.1525/aa.1964.66.3.02a00850

Toledo, V. M. 1992. What is ethnoecology? Origins, scope and implications of a rising discipline. *Etnoecologica*, 1: 5–23.

Warren, D. M., Slikerveer, L. J. and Brokensha, D. (eds.) 1995. *The Cultural Dimension of Development: Indigenous Knowledge Systems*. London, Intermediate Technology Publication.

White, Jr., L. 1967. The historical roots of our ecologic crisis. *Science*, vol. 155. no. 3767: 1203–1207.

Part I

From Practice to Principles

Methods and Challenges for Decolonized Knowledge Co-production (DKC)

2

The Progression from Collaboration to Co-production: Case Studies from Alaska

HENRY P. HUNTINGTON, GEORGE NOONGWOOK, ANNE K. SALOMON
AND NICK M. TANAPE, SR

Introduction

Over the past two decades, the idea of using local and traditional knowledge (LTK) in research and management of the environment has moved from the fringes towards the centre. Discussions on the topic are now moving from the question, 'Why should we do it?' towards 'How do we do it?' In part, this trend is about making explicit what had been done in practice for far longer. Early explorers, adventurers and scientists travelling to the Arctic drew on the expertise of Arctic Indigenous peoples for survival, navigation and insight. They rarely gave credit to their instructors, though there were some exceptions. Many scientists in the late twentieth century also learned from local residents, often giving recognition in the 'Acknowledgements' section of a paper or report.

The increase in attention to LTK (also under various other names, such as Indigenous science, or traditional ecological knowledge) brought greater awareness among Indigenous peoples of the significance of their knowledge and experience beyond their own communities. Today, the holders of LTK are included in many studies as authors of papers and reports, reflecting a greater recognition of their intellectual roles and rights by academic researchers. These are welcome developments, marking the progression from 'local colour' to 'informant' to 'participant' to 'collaborator'.

In addition to greater recognition, another trend has been taking place in studies that involve LTK. This trend is one of content, in the evolution of the ways in which LTK is used and, more importantly, in the ways that LTK holders and academic researchers work together. (We use the terms 'LTK holders' and 'academic researchers' as shorthand to distinguish two ends of a spectrum, recognizing that there are many individuals who are both LTK holders as well as academics.) Here, the progression is from collaboration, which can describe many modes of working together, to knowledge co-production, which is a specific paradigm under which 'new' knowledge is generated by the combined intellectual efforts of all involved. The term 'knowledge co-production' typically describes a collaborative approach throughout the entire research process (Lemos and Morehouse, 2005; Meadow et al., 2015), but here we include examples in which a co-production approach developed as a project evolved.

Recognizing that projects and collaborations evolve, in this chapter, we distinguish between the 'documentation of knowledge', which makes accessible to others what is already known to some; the 'connection of knowledge', which links what is already known by different people or in different fields; and the 'co-production of knowledge', in which new knowledge is generated upon the understanding and insights of two or more people, groups or perspectives. We draw on two studies that we have been part of in Alaska: the Bering Sea and Bidarki Projects, which give examples of these three modes and of the conditions that foster moving from one to the next.

The Studies

The Bering Sea Project was a joint endeavour by the North Pacific Research Board and the US National Science Foundation, conducted from 2008–2013. One component of this large project focused on local and traditional knowledge (LTK, the term used by the North Pacific Research Board). In this part of the study, we worked with five Bering Sea communities: Akutan, St. Paul, Togiak, Emmonak and Savoonga, to (1) document subsistence harvests; (2) document LTK about the Bering Sea ecosystem; and (3) analyse our findings in combination with those of other parts of the overall project. In this chapter, we emphasize part (3) as the realm of co-production of knowledge.

The purpose of the whole project was to better understand the variability of the ecosystem and the connections among climate, physical oceanography, nutrients, plankton, fish, seabirds, whales and humans (Wiese et al., 2012; Van Pelt, 2015). The Bering Sea accounts for about half of US fish catches, worth some US$2 billion per year, and is also experiencing rapid environmental change. Understanding what changes are likely in the future is vital for sound fisheries management, for ensuring continued subsistence harvests, and for conserving the species, habitats and cultures of the region.

The five communities involved in the LTK component were approached by social scientists working on the research proposal and their participation was confirmed once the research grant was secured. The basic parameters of the LTK research such as harvest surveys and LTK documentation were laid out by the social scientists, including Jim Fall and colleagues from the Alaska Department of Fish and Game, Division of Subsistence; Jennifer Sepez from the National Oceanic

Figure 2.1 Participants at one of the annual meetings of principal investigators of the Bering Sea Project, including George Noongwook and Henry Huntington.
Photo by Glenn Aronwits

and Atmospheric Administration; Eugene Hunn from the University of Washington; Henry Huntington from Huntington Consulting; and researchers who are also community members, such as George Noongwook from Savoonga and Phil Zavadil from St. Paul. The research plan was carried out with little modification, as both components were fairly straightforward and had community support. The process of connecting our findings with other parts of the Bering Sea Project took place later and is the main topic of this paper and is therefore discussed in detail below.

A group of researchers and community members, some of whom filled both roles, managed the LTK component. This group helped plan the overall research (parts (1) and (2)) and oversaw the post-fieldwork effort of comparing results from the five communities and connecting our findings with those from other parts of the Bering Sea Project (part (3)). A major contributor to part (3) was the participation by community members in the annual principal investigators' meetings of the Bering Sea Project (Fig. 2.1). As was the case for all components of the project, the connections and relationships among individuals improved from year to year, leading to greater understanding of everyone else's work, better communication, and eventually, to connecting and co-producing knowledge in ways that could not have

Figure 2.2 A *bidarki* in its natural habitat during low tide on the lower Kenai Peninsula, Alaska. This individual is about 8 cm long. [A black and white version of this figure will appear in some formats. For the colour version, please refer to the plate section.]
Photo by Henry Huntington

been anticipated at the outset of the project. In other words, space was created for the social interactions that led to intellectual collaborations.

The Bidarki Project was conducted in the 2000s under the leadership of Anne Salomon, then a graduate student at the University of Washington and now a professor at Simon Fraser University in Canada. She began by looking at a topic that the Alutiiq communities of Nanwalek and Port Graham had identified as a concern: the decline of the *bidarki*, the local name for the black leather chiton (*Katharina tunicata*), an intertidal invertebrate and culturally important source of subsistence food (Fig. 2.2). This invertebrate was also well known as a keystone grazer and Anne was fascinated by the potential trophic cascade that might have been triggered by its decline. While motivated by both conservation concern and ecological fascination, Anne's project began as a standard marine ecology study, using standard methods to compare *bidarki* density and growth at several sites across a range of distances from the two communities. The project was funded only by graduate fellowships at the time and local community members volunteered on the project, sharing their time, boats and knowledge out in the field while collecting ecological data.

Anne quickly discovered how much the community members had to say about the ecosystem as a whole and *bidarkis* in particular (Huntington et al., 2011). Although distant from her training, it became glaringly obvious to Anne that learning from and working with local knowledge holders was an essential way of understanding both the ecological and social processes of the entire system, past and present. So, after that first summer of collaborative field work, Anne and the people of Port Graham and Nanwalek collaborated in the drafting of a research proposal to identify the causes and consequences of *bidarki* declines. It was funded through the Gulf of Alaska Ecosystem and Monitoring (GEM) programme and the Exxon Valdez Oil Spill Trustee Council, responsible for administering funds for research of the ecosystem damaged by the spill. Although Anne and the tribes had specifically added an LTK component to the proposal, they both needed guidance on how best to collect and integrate this data with the ecological data.

At the same time, the Trustee Council had begun to emphasize the role of LTK in its work, which included a grant to Chugach Regional Resources Commission (CRRC), a tribally authorized organization active in the spill region of the southcentral Alaska coast. Henry was hired under contract to CRRC to support LTK studies and was subsequently contacted by Anne to see if he could help her with the newly expanded and funded Bidarki project, which now included a more substantive LTK component. With some additional funding from the Trustee Council, Anne and Henry conducted interviews with residents of Port Graham and Nanwalek and held an Elders' lunch as a setting for further discussions. They also took part in a workshop in Port Graham with a wider scope, funded by the Trustee Council and organized by CRRC, which provided an additional opportunity to discuss the findings of the Bidarki Project with local residents as well as other visiting scientists. Lastly, research update meetings with the communities at the end of every summer and annual middle school intertidal science field trips provided additional spaces for knowledge exchange. In addition to these formal dialogues, countless informal discussions on the docks, in the field and around the kitchen table informed everyone's collective understanding of the causes and consequences of *bidarki* declines.

Ethics

In both cases, the LTK research was conducted in accordance with the principles of free, prior and informed consent. The tribal councils in each community, which, in Alaska, are the Indigenous governments responsible for tribal affairs, were asked if they would agree to participate in the project. Once their consent was obtained, we made plans for specific research activities. Interviews were conducted with groups or with individuals. In each case, the researchers described the overall project and the LTK component, along with the methods that would be used and the ways in which the information would be reviewed by interview participants. Participants were assured that they were free to leave the study at any point. The Bering Sea Project paid participants for their time. In most cases, the researchers included local residents as well as visiting academically trained researchers in the fieldwork. In St. Paul, the research was carried out entirely by local residents who have previously undertaken several such projects on their own.

When the interviews were completed, the researchers prepared notes in the form of a narrative describing what participants had expressed about the ecosystem, the species, human activities, and so on. This set of notes was sent to participants for their review and, where possible, was reviewed together with one of the researchers. Here, the role of the local researcher was often critical, as budget constraints often limited the ability of the visiting researcher to make an additional trip to the community in question.

In the Bering Sea Project, once the notes were corrected and approved, they were submitted to the central data archive for the entire Bering Sea Project (http://beringsea.eol.ucar.edu/), in accordance with the terms of the grant. Metadata was also provided in accordance with best practices in data management. The main findings of the LTK component and other parts of the Bering Sea Project were summarized in 'Headlines' reports intended for a wide audience, including the communities participating in the LTK effort, members of the public throughout the region and Alaska, non-specialists or those with expertise in other fields, and so on. This part of the project came at the end, as the many scientific papers were being completed. Each scientific paper had one headline report to help make sure that the results were communicated to all who were involved as well as to the wider audience of interested people. These reports are available at the Bering Sea Project website (www.nprb.org/bering-sea-project) and several printed copies were sent to each of the participating communities. In the Bidarki Project, the results were shared with the communities at annual presentations and by the collaborative production of a book (Salomon et al., 2011) written for a wide audience and distributed in Port Graham and Nanwalek. The results of both projects have also been presented at several conferences and at regional meetings of tribes and Alaska rural residents.

Approaches

The LTK interviews in the Bering Sea Project were mainly conducted using the semi-directive interview (Huntington, 1998), although one of the five communities, St. Paul, decided to use a more formal questionnaire. In this method, the researchers introduce topics for discussion but allow those discussions to follow

Figure 2.3 Interview in Savoonga to document local and traditional knowledge about the Bering Sea ecosystem. [A black and white version of this figure will appear in some formats. For the colour version, please refer to the plate section.]
Photo by Henry Huntington

the trains of thought and connections that the participants see, rather than sticking to a formal questionnaire or other structured means of guiding the discussion. In this way, the local perspective is allowed to guide the discussion, often leading to insights into ecosystem interactions and connections that might otherwise be missed (Fig. 2.3). The researchers may steer the conversation at times, or introduce new topics when appropriate to keep the conversation going and to make sure that all relevant topics are covered at one point or another. In addition to the semi-directive interviews, the Bidarki Project researchers also used maps and a questionnaire to document the location and magnitude of *bidarki* subsistence harvest effort, the sizes of *bidarkis* currently harvested versus the sizes harvested in the past, and how sizes in the past varied with the age of the harvester.

In the Bering Sea Project, making connections with other parts of the overall project took place primarily at the annual researchers' meetings. At these events, the hundred or so researchers involved in the project presented their activities and findings, and there was ample time for discussion and collaboration. Local as well as visiting researchers represented the LTK group in order to maximize opportunities to establish connections and learn from one another. This level of participation was costly, but well worth the investment. We were also able to hold meetings of just the LTK group at the same time, taking advantage of the opportunity to be together at least once a year.

Many of the connections were largely a matter of chance. We tried to create the right conditions for such connections by paying attention to the interests of other project components, by inviting other researchers to visit one or more of the five communities, and by taking part in the various small-group discussions held at the annual researchers' meetings. One difficulty we encountered was that our research started a year later than most other components because we spent the first year obtaining tribal council permissions and otherwise getting our project set up. As a result, our findings came after many other field efforts were either completed or established in ways that could not be changed despite newfound opportunities to work together. In addition, the resolution of our findings in time and space often did not match those of other project components. In retrospect, it would have been good to have had a chance to return to each village to discuss their LTK again in light of what we learned from our fellow researchers.

In the Bidarki Project, our main effort centred on the project itself, with additional strands of inquiry being pursued as they became relevant. What had started as an ecological field study also became an LTK study as Anne learned more and more from expert local residents such as Nick Tanape, Sr. from Nanwalek, who became our co-author on both our scientific article (Salomon et al., 2007) and our book (Salomon et al., 2011). Beyond sharing ecological knowledge, Nick and others shared their insights into the past: past subsistence ways, past socio-political events that altered the way people harvested resources and past cultural practices, be they Alutiiq dances, many of which reflected marine animal behaviour, or the rituals surrounding hunting practices. Consequently, what started as an ecological project transformed into a social-ecological project, one that aimed to capture the changes in interactions between social systems, ecosystems and management systems through time that led to the current *bidarki* decline (Salomon et al., 2007, 2011) Notably, all of the Elders interviewed emphasized the importance of their language, *Sugestun*, in recalling, understanding and teaching LTK to the younger generation so that this knowledge can be used to help solve many of the environmental challenges that Indigenous communities face today. This common call prompted the LTK presented in our book to be translated into *Sugestun* as a teaching tool for the youth of these and other Chugach tribes.

One important (though hardly unique) aspect of the Bidarki Project was its genesis as a field project, so that Anne, Nick and others had the experience of working together at the study sites, seeing the same things and talking about how they understood what they were seeing (Fig. 2.4; Huntington et al., 2011). This approach differs from many LTK projects that depend primarily or exclusively on

Figure 2.4 Nick Tanape, Sr. (left) and Anne Salomon (right) in the field with Dave Glahn (centre) during the Bidarki Project. Note the tools Nick is holding, used for conducting standard ecological research in the intertidal zone.
Photo courtesy of Anne Salomon

interviews. In those cases, the interviewer may not fully understand what the interviewee is describing, creating the potential for misunderstanding or misinterpretation. While those flaws are not eliminated by spending time in the field together, having a solid common reference point was invaluable to the co-production of knowledge and the development of our paper and book. Not only did it allow for the production of new knowledge, but it also established the prerequisite of trust and equity upon which our relationship and shared knowledge was based.

In addition to the ecological and LTK parts of the study, other topics were introduced as needed. For example, the analysis of archaeological remains or the compilation of the story of settlement patterns in the region were done as it became apparent that we needed a deeper historical context in which to interpret our findings. Where appropriate, we contacted other scientists to gain access to their expertise and data or delved into archives or historical records to supplement the memories and knowledge of local residents. As the project grew, or more specifically, as our ambitions grew to include the production of a book that would capture the spirit of the endeavour, we sought additional funding from various sources as well as a publisher with a similar vision. This effort required considerable patience and persistence, primarily on Anne's part, but in the end resulted in a book that won an Alaska Library Association award as well as local acclaim (Salomon et al., 2011).

Outcomes

Four papers and the book that resulted from these projects illustrate the three types of contributions to knowledge described in the Introduction to this chapter. One paper from the Bering Sea Project 'presented the knowledge we documented' in our LTK interviews (Huntington et al., 2013a). The purpose of this paper was to make available the knowledge held by hunters, fishers and gatherers in the five villages, and to provide some discussion about how that knowledge matched or did not match understanding and insight from other research in the region. We reported changes in species abundance, changes in distribution of species, the locations of ecological hotspots and insight into the health of the ecosystem as a whole. Overall, our findings were largely consistent with what other scientists had found and with what other community-based projects had shown. We felt that this paper was a valuable contribution, allowing the communities' LTK to speak for itself, and drawing further attention to the depth and breadth of their insight into the Bering Sea ecosystem. This paper was a straightforward presentation of the results of our fieldwork and was an expected outcome of the LTK effort. The discussion of how the LTK results fit in with findings from other parts of the Bering Sea Project or from previous studies was aided by the connections we had made with other researchers throughout the project, but again this was a relatively modest level of interaction and could have happened in other ways as well.

Another Bering Sea paper 'compared the results of three approaches' to documenting subsistence use areas (Huntington et al., 2013b). One community-based study documented individual hunting trips shortly after they happened over the course of one or more years. They were thus able to show where people hunted and fished and how intensively those areas were used over time and by season. This approach also allowed them to show variability from one year to the next. Another community-based study documented lifetime subsistence use areas via interviews. These results showed the vast areas that may be used in one year or another, depending on conditions and need. The third approach was a novel one, looking not at the areas where hunters and fishers travel, but at the areas where the animals go according to satellite telemetry and mark-recapture methods. We called these the 'calorie-sheds' in recognition of the fact that they showed where the food came from before it arrived on the dinner table. The calorie-sheds cover vast areas including, for Savoonga, much of the Bering, Chukchi and Beaufort seas as well as the Gulf of Alaska.

From the community perspective, it is important to know how much of the sea they draw upon for their well-being. Decisions and activities anywhere in their calorie-shed can affect them, not just those in the areas where they hunt and fish. To delineate the calorie-sheds, we connected existing information about subsistence harvests, such as which species were most commonly harvested in a given village, with available data on the range of the species or stocks that they were harvesting, recognizing that such data were not available for many species. This paper was valuable in showing how LTK data about harvests and local ecology can be connected with data from scientific methods to give us a new way of looking at a system. The novel way of understanding the system was made possible by the interactions among community members, social scientists and biological oceanographers, who recognized that each group held part of the picture, but no one had all parts until we worked together.

The third Bering Sea paper took us 'towards the realm of co-production'. Walrus hunters in Savoonga described large-scale patterns in ice formation and movements during winter, and explained how those patterns affect spring break-up and the success of spring walrus hunting. At one of the annual researchers' meetings, Jinlun Zhang, a sea ice modeller, showed an animation from remote sensing and models of ice movement through two winters, which exactly followed the pattern described by the hunters. George Noongwook, who was present at this and all other researchers' meetings during the project, was able to discuss the convergence with Jinlun, and we soon recruited Nick Bond, a climate analyst and modeller, to address the role of wind and to develop a statistical model. From this initial connection made by looking at Jinlun's animation of ice, we began an analysis of the influence of ice concentration and wind conditions on walrus hunting effort and success (Huntington et al., 2013c).

Building on the extensive and detailed data that were available on weather, ice and the walrus harvest (the last coming from Brad Benter and Jonathan Snyder at the US Fish and Wildlife Service), we were able to focus our efforts on the conditions that affect daily hunting success. We used a generalized additive model (GAM) to conduct the statistical analysis, which established that hunting effort, that is, the number of trips taken to hunt, can explain most of the variation in hunting success, which is defined as the number of walrus harvested. Variation in wind and ice conditions, however, explained only around 20 per cent of the day to day variability in harvest, leading us to conclude that other factors are likely involved as well, including social factors that affect participation in hunting. In carrying out this effort, we built on the observations of Savoonga hunters (Fig. 2.5)

Figure 2.5 George Noongwook driving his boat along the north coast of St. Lawrence Island: an experienced hunter in his element.
Photo by Henry Huntington

and added other data and analytical and modelling techniques. Most of this effort was sequential in nature, rather than having repeated feedbacks among the various parts of the analysis. Nonetheless, the new knowledge we generated about the relationship between physical conditions and hunting outcomes would not have been possible without the strong intellectual contributions from all who took part, a hallmark of knowledge co-production.

The *bidarki* paper and book (Salomon et al., 2007, 2011) are the culmination of the journey to knowledge co-production. Anne's field studies found that *bidarki* density increased with distance from the community, suggesting that harvest played a role. She and her fellow researchers also found that the presence of sea otters had a similar effect to the presence of human harvesters. Consequently, both top predators were responsible for the spatial variation in *bidarki* numbers and size. These and related findings informed us about *bidarkis* in the present day but revealed little in the way of what had actually caused their decline through time, which was the ultimate question of the project. The LTK interviews began to add the much needed historical context, making locally held knowledge available to more people (such as Anne and Henry) as we documented major events in the history of both communities and significant ecological changes, such as the return of sea otters to the area following their decimation during the fur trading era of the nineteenth century.

A new picture emerged, in which the decline of *bidarkis* was just the latest in a series of invertebrate declines dating back to the return of sea otters in the 1950s and exacerbated by changes in patterns of human use. These included connecting Port Graham and Nanwalek to the region's electrical grid in the early 1970s, which allowed people to buy freezers and store surplus harvests. The Exxon Valdez oil spill caused ecological devastation in many areas (though not in these two communities) but also brought a chance to earn money and purchase boats with outboard motors. This made it easier for *bidarki* gatherers to travel farther and harvest more. Looking deeper into the historical and archaeological record extended the basic idea of a concentrated human footprint and substantial change in harvest patterns over long periods. In short, a previously unknown story developed out of our collective effort, one that could not have emerged without both streams of inquiry and knowledge. The scientific paper tells the story in academic terms, focusing on ecology and ecosystem interactions. The book tells the story in human terms, including the implications of our findings for local use and management and the future of the Alutiiq communities that hold the *bidarki* in such high regard.

Conclusion

The significance of all four papers and the book is that they demonstrate the connections between Alaska Native communities and their ecosystem, and provide some insight into how changes in climate, environment and human activity may affect traditional as well as modern activities. The documentation of subsistence use areas and calorie-sheds, for example, helps show why Savoonga is concerned about shipping in the Bering Strait as well as oil development in the Chukchi and Beaufort Seas. The analysis of how wind and ice affect walrus hunting provides insight into how Savoonga might be affected by further climate change and resulting impacts on wind and ice. It also shows that other factors are important, including the resilience and innovation of Savoonga's hunters, who are used to dealing with variability and are likely to continue to find ways to hunt successfully even as conditions change. The full story of the *bidarki* helps identify specific actions local residents can take to shape the future of their ecosystem, as well as sources of impacts that are beyond their immediate control, such as commercial fishing in nearby waters and climate change.

The papers and book also show how co-production of knowledge often rests on the documentation of knowledge and the connection of different forms of knowledge, as ways of creating relationships and the foundation for co-producing new knowledge that goes beyond what was previously known. Not all efforts to document LTK continue to the co-production stage, but few co-production efforts can be undertaken without making the community's knowledge available in one way or another. Documentation for public distribution may not be suitable in all cases, but a strong community role is essential to create a joint effort and not simply to allow others to document and apply the results of LTK. We also recognize that the ideal form of knowledge co-production starts at the very beginning, with a joint effort to identify the scope and aims of the project. Our examples show that projects can also evolve in the direction of knowledge co-production once they are underway, if the collaborations are built on mutual respect and shared contribution.

Projects that document or compare LTK should not be dismissed as incomplete or inadequate. These are important contributions to research and often have lasting and meaningful benefits for the individuals and communities involved. Indeed, these stages on the path to co-production can have valuable personal outcomes, such as a greater recognition of the importance of one's knowledge or of the enjoyment to be had in working with others on a topic near and dear to one's heart. Getting more people interested and involved in this kind of work can be a lasting benefit. In neither the Bering Sea Project nor the Bidarki Project did we aim for knowledge co-production from the beginning. If we had been held to that standard at the outset, we might not have begun for fear that we were taking on too much. Instead, the projects evolved towards co-production as we proceeded, as trust was built and relationships were created among the participants, and as people began to realize that more was possible. This progression is most evident in the Bidarki Project, for which more funding had to be found and research partners added as it developed. In the Bering Sea Project, we recognized the possibility of grander outcomes, and had built some room into our budget to support collaborations that would emerge along the way.

In short, it is difficult or impossible to know at the beginning of a project or collaborative relationship how it will grow and whether knowledge co-production is a plausible outcome. Instead, it is valuable to know that the possibility exists, so that local residents and visiting academics can be open to seeing the potential when it begins to appear. What can be done at the beginning is to acknowledge the importance of working together to achieve any of the aims described here and in other chapters of this book. This is the essential foundation of success, from deciding the scope and aims of a project to the accurate documentation of knowledge and then to the ability to compare knowledge across areas or disciplines and finally to the development of true knowledge co-production. Each collaborative project is a step in the direction of mutual respect and a sense of intellectual equality for all who are involved. The more efforts that begin, the more

Figure 2.6 Anne Salomon and Nick Tanape, Sr. with a copy of *Imam Cimiucia*, the result of knowledge co-production in Nanwalek and Port Graham.
Photo courtesy of Anne Salomon

experience we all gain and the greater the likelihood that some of these relationships and studies will progress all the way to the co-production of knowledge (Fig. 2.6).

Acknowledgements

We are grateful to the North Pacific Research Board (NPRB), the Gulf Ecosystem and Monitoring (GEM) Programme and the Exxon Valdez Oil Spill Trustee Council for funding our work, including the opportunities to interact with other

Bering Sea Project and Gulf Ecosystem Monitoring and Research Project scientists. We are also grateful to all the participants in our study, whose time and willingness to share knowledge are the foundation of our work, and to our many collaborators, without whom none of these projects would have been completed. Finally, we are grateful to Igor Krupnik, Shari Fox Gearheard, Doug Nakashima and Marie Roué for encouragement and constructive comments on earlier drafts. Any mistakes that remain are of course ours alone.

References

Huntington, H. P. 1998. Observations on the utility of the semi-directive interview for documenting traditional ecological knowledge. *Arctic*, 51(3): 237–242. https://doi .org/10.14430/arctic1065

Huntington, H. P., Gearheard, S., Mahoney, A. and Salomon, A. K. 2011. Integrating traditional and scientific knowledge through collaborative natural science field research: identifying elements for success. *Arctic*, 64(4): 437–445. https://doi.org/ 10.14430/arctic4143

Huntington, H. P., Braem, N. M., Brown, C. L., Hunn, E., Krieg, T. M., Lestenkof, P., Noongwook, G., Sepez, J., Sigler, M. F., Wiese, F. K. and Zavadil, P. 2013a. Local and traditional knowledge regarding the Bering Sea ecosystem: Selected results from five Indigenous communities. *Deep-Sea Research II*, 94: 323–332. https://doi.org/10 .1016/j.dsr2.2013.04.025

Huntington, H. P., Ortiz, I., Noongwook, G., Fidel, M., Alessa, L., Kliskey, A., Childers, D., Morse, M. and Beaty, J. 2013b. Mapping human interaction with the Bering Sea ecosystem: Comparing seasonal use areas, lifetime use areas, and "calorie-sheds". *Deep-Sea Research II*, 94: 292–300. https://doi.org/10.1016/j.dsr2.2013.03.015

Huntington, H. P., Noongwook, G., Bond, N. A., Benter, B., Snyder, J. A. and Zhang, J. 2013c. The influence of wind and ice on spring walrus hunting success on St. Lawrence Island, Alaska. *Deep-Sea Research II*, 94: 312–322. https://doi.org/10 .1016/j.dsr2.2013.03.016

Lemos, M. C. and Morehouse, B. J. 2005. The co-production of science and policy in integrated climate assessments. *Global Environmental Change*, 15(1): 57–68. https:// doi.org/10.1016/j.gloenvcha.2004.09.004

Meadow, A. M., Ferguson, D. B., Guido, Z., Horangic, A. and Owen, G. 2015. Moving toward the deliberate co-production of climate science knowledge. *Weather, Climate and Society*, 7: 179–191. https://doi.org/10.1175/WCAS-D-14-00050.1

Salomon, A. K., Tanape Sr., N. M. and Huntington, H. P. 2007. Serial depletion of marine invertebrates leads to the decline of a strongly interacting grazer. *Ecological Applications*, 17(6): 1752–1770. https://doi.org/10.1890/06-1369.1

Salomon, A., Huntington, H. P. and Tanape Sr., N. 2011. *Imam Cimiucia: Our Changing Sea*. Fairbanks, AK: Alaska Sea Grant.

Van Pelt, T. I. (ed.) 2015. *The Bering Sea Project: Understanding Ecosystem Processes in the Bering Sea*. Anchorage, AK: North Pacific Research Board. https://doi.org/10 .13140/RG.2.1.1668.2482

Wiese, F. K., Wiseman, W. J. and Van Pelt, T. I. 2012. Bering Sea linkages. *Deep-Sea Research II*, 65–70: 2–5. https://doi.org/10.1016/j.dsr2.2012.03.001

3

Learning about Sea Ice from the Kiŋikmiut: A Decade of Ice Seasons at Wales, Alaska, 2006-2016

HAJO EICKEN, IGOR KRUPNIK, WINTON WEYAPUK, JR.
AND MATTHEW L. DRUCKENMILLER

Dedicated to the memory of our late partner and friend, Winton (Utuktaaq) Weyapuk, Jr.
(1950–2016)

Over the past decade, the story of the decline of Arctic summer sea ice has drawn widespread attention. Overwhelming evidence indicates that Arctic sea ice is rapidly changing, with an ongoing Arctic-wide decline since the 1970s and accelerating ice loss since the year 2000. Particularly large reductions in summer ice extent began with the then record-setting minimum in 2007 (Stroeve et al., 2008) and continued into the next decade. The thirteen lowest summer ice extents, since the start of satellite observations in 1979, were all recorded between 2007 and 2019 (National Snow and Ice Data Center, 2019), with the lowest in 2012. Equally remarkably, the total volume of Arctic sea ice has diminished by more than 75 per cent over this same time period, from 1979 until today (PIOMAS, 2020), which is largely a result of the loss of the Arctic's old multi-year ice floes.

With this dramatic circumpolar decline, some of the fastest sea ice loss occurred in the North Pacific-Western Arctic region, particularly along the Alaskan coast. Although sea ice loss has been recorded in all months, satellite observations show the fastest rate in late summer and autumn. Since 1979, the annual average Arctic sea ice extent has decreased at a rate of 3.5 per cent per decade, while regional sea ice melt along the Alaskan coast, in the Beaufort and Chukchi Seas, exceed the Arctic average, at 4.1 per cent and 4.7 per cent per decade, respectively (Taylor et al., 2017).

The sea ice melt season – defined as the number of days between spring melt onset and fall freeze-up – has lengthened Arctic-wide by at least five days per decade since 1979. Again, some of the largest observed changes have been recorded along Alaska's northern and western coasts, lengthening the melt season by 20–30 days per decade and increasing the overall annual number of ice-free days by more than ninety, that is, by a full three months (Taylor et al., 2017).

Summer sea ice retreat along Alaska's coasts has led to longer open water seasons, and more frequent and violent storms have made the Alaskan coastline more vulnerable to erosion.

Yet, the long-term passive-microwave satellite data critical to these observations (e.g., Comiso, 2017) have their inherent limitations. As the satellite orbits the Earth, each measurement, a so-called pixel, is pieced together into a large mosaic of the Arctic. A pixel covers roughly 25×25 km. The entire width of Bering Strait, at 82 km, and a portion of the adjacent area are thus covered by just four such pixels. This coarse scale is sufficient for a general study of how Arctic sea ice fluctuates and helps regulate Earth's climate. However, such data are of less value if we want to learn about how marine mammals utilize different ice types or to get practical information for coastal communities that rely on ice that is firmly attached to land (shorefast) and drifting ice for hunting, transportation and other uses.

In a changing North that also experiences increasing shipping traffic and industrial activities, demand is great for detailed information on the characteristics of the ice cover and its seasonal waxing and waning. Planning and oversight associated with various activities require both broad and in-depth understanding of the ice, typically for a particular location. For this, many specific questions have to be answered. In order to assess potential impacts in the context of increasing human activities in the Arctic, information from science is often not enough.

Perspectives of Scientists and Indigenous Users on Sea Ice

Scientists often approach research questions by breaking down the task into smaller, manageable pieces, each studied by a group of experts from a specific discipline. While this approach is well proven, it has its limitations in addressing broader cross-disciplinary issues, such as the impacts of changing ice on the environment, marine ecosystems and its human uses.

A very different way of studying the ice is derived from its practical use: as a provider of food or as a platform for travel and other activities. In Alaskan coastal communities, knowledge of sea ice dynamics by local experts is based largely on close, repeated observations, building off a vast body of experiential records that is continually tested, reaffirmed or modified by time spent observing the ice from shore, boats and on the ice itself. A number of studies have documented and discussed Inuit, Iñupiaq and Yupik sea ice expertise in great detail (Nelson, 1969; Krupnik and Jolly, 2002; George et al., 2004; Oozeva et al., 2004; Gearheard et al., 2006, 2013; Laidler, 2006; Laidler and Elee, 2008; Laidler and Ikummaq, 2008; Krupnik et al., 2010a). Thanks to this body of work, it is obvious that Indigenous communities across the Arctic are aware of the many facets of rapid environmental change that are affecting their home regions and continue to adapt to such change.

People who are engaged in daily activities on ice and on the land and who rely upon traditional knowledge and close monitoring of various aspects of the environment are often the first to detect changes in a particular location. Collaboratively exploring the profound differences between how communities of researchers, engineers and Indigenous users collect, assess, store and disseminate sea ice knowledge can lead to new knowledge and understanding and serve as an eye-opener to modern scholarship on Arctic change.

Building a Network of Local Observers in Alaskan Communities

The process of sharing knowledge between Indigenous residents and visiting explorers, scientists and traders might have started during the earliest days of contact in the Arctic. 'Systematic' collection of Indigenous knowledge about sea ice and ice impact on northern marine mammals did not take place until much later, mostly in the 1960s and 1970s (see Krupnik, Chapter 5). By the late 1990s, as Indigenous residents started to voice their concerns about the warming of the Arctic, a new field of collaboration and knowledge co-production was born, that focused on documenting Indigenous observations of Arctic sea ice and climate change.

The first pilot studies at the turn of the century (McDonald et al., 1997; Krupnik, 2000, 2002; Fox, 2002; Oozeva et al., 2004) made it clear that our understanding of the changing Arctic could not advance without the systematic participation of Indigenous experts from polar communities as, for example, long-term ice and weather observers in their home areas. A determined effort was launched in 2006 to coincide with the International Polar Year (IPY) 2007–2008 as a collaborative venture between sea ice and cultural researchers on the one hand and Alaskan Native knowledge experts from several communities on the other. This took the shape of two IPY projects, 'SIKU' (Sea Ice Knowledge and Use) and 'SIZONet' (Sea Ice Zone Observing Network: Krupnik, 2009; Eicken, 2010; Eicken et al., 2014). As part of these projects, Alaska Native hunters agreed to keep daily or near-daily notes on local weather and ice conditions as relevant to their activities on shorefast ice or among drifting ice.

In contrast with standard sea ice logs, such as those compiled by ship-based observers (Worby and Eicken, 2009), this new observation programme was not constrained by a standardized observation protocol. Instead, observations were guided by sea ice use in a particular location. Local observers were asked to note general ice conditions and key weather variables when relevant, including temperature and wind speed and direction, obtained from residential weather stations. More specifically, they focused on the timing of key events in the annual ice cycle, such as the appearance of the first slush or drifting ice when the ice becomes safe for travel and the timing of ice breakup. They were encouraged to report any local details they deemed important, such as those having to do with the ice environment, subsistence activities, sea ice travel, wildlife associated with different forms of ice and community events. The observers were also encouraged to use terms from their Indigenous languages, specific local place names and local forecasting indicators, as well as to reference their personal experiences and memories of other community members when relevant.

The first Alaska Native observer to join this new programme was the late Leonard Apangalook, Sr., a Yupik hunter from the community of Gambell on St. Lawrence Island, who started his daily recording in the spring of 2006 which was continued after September 2008 by his brother, Paul Apangalook (1951–2018; Krupnik et al., 2010b). In fall 2006, two Iñupiat experts, Winton Weyapuk, Jr. from Wales and Joe Leavitt from Utqiaġvik (Barrow), signed on as SIZONet observers and subsequently both continued their service for almost a full decade until Weyapuk's passing in November 2016. During peak activities from 2008 to 2010, additional observations were conducted in several rural communities in Alaska, such as Shaktoolik and Toksook Bay; and the Russian Chukotka on the Siberian side of Bering Strait including Uelen, Yanrakynnot, New Chaplino, Provideniya and Sireniki (Bogoslovskaya and Krupnik, 2013; Fig. 3.1).

The observers' logs provided a daily record of conditions throughout the full ice cycle, usually from September/October to May/June of the next year. While the focus of records was on sea ice and weather conditions and ice-associated activities, the observers also described how the changing sea ice affected subsistence and community life. They noted potential hazards, old and new, and identified specific weather or ice events or wildlife sightings. Observers also lent cultural and historical context by relaying discussions with, or comments by, Elders and other hunters, and by referencing stories that had been passed down through generations. Sea ice terms were often noted in local languages. All observational data were initially uploaded into a database at the Geophysical Institute, University of Alaska Fairbanks. Through collaboration with the Exchange for Local Observations and Knowledge of the Arctic (ELOKA) project, a dedicated long-term archival database with an observer interface was put in place (Eicken et al., 2014).

Enter Winton Weyapuk, Jr.

From fall 2006 until his untimely passing in November 2016, Winton (Utuktaaq) Weyapuk, Jr. from Wales (Kiŋigin) was the most dedicated and trusted local partner

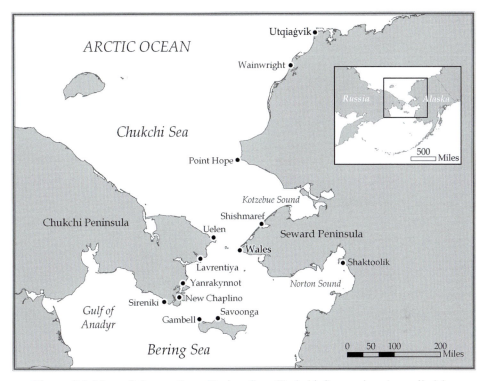

Figure 3.1 Map of the northern Bering Sea–Chukchi Sea region (compiled by Matthew L. Druckenmiller)

in a decade-long effort. Weyapuk, a hunter and whaling captain, was a born naturalist, but also an educator, community leader and strong proponent of his native Kiŋigmiut dialect of the Iñupiaq language, with a keen understanding of the role of partnering with scientists in preserving his people's knowledge and heritage (Krupnik, 2017; Fig. 3.2). A thoughtful and humble man, he was interested in the links between the changing ice and weather, the impact on marine mammals and subsistence hunting, and the role of Indigenous knowledge in supporting sustainability and the spiritual wellbeing of his small community with a population of 150. For these and many other qualities, he was a remarkable partner in a joint co-production effort that lasted over ten years (see also Krupnik, Chapter 5).

Observations of a Changing Ice Environment

What did hunters and Indigenous ice experts like Weyapuk observe at coastal Alaskan sites, specifically at Wales, during these years? Here, we report on some aspects of Weyapuk's work for his community of Wales located at the westernmost point of the North American continent, Cape Prince of Wales

Figure 3.2 Winton Weyapuk, Jr. on the ice near Wales, Alaska.
Photo by Chris Petrich, 2008

(65°35′47″N, 168°05′05″W) and facing the Asian mainland across the 82 km (51 mile) wide Bering Strait. Weyapuk's near-daily ice observations from fall of 2006 through spring of 2016 generated an unparalleled trove of data. Observations by Weyapuk and satellite data for 2006–2007 are summarized in Fig. 3.3.

According to the weather station at Tin City located less than 10 km (7 miles) from Wales, the first day with freezing temperatures in fall 2006 was 6 October. Temperatures fell continuously below freezing from 4 November onwards. In his daily logs, Weyapuk noted between 10 and 15 November the first occurrence of an ice slush berm (Inupiaq: *qaimġuq* – see Fig. 3.4) accumulating through wave action and wind driving slush ice (Inupiaq: *qinuliaq*) ashore. This first ice is very important for coastal communities as it usually buffers the coast from ocean waves. In places like Shishmaref, 120 km (75 miles) from Wales along the coast, early formation of such a slush berm offers protection from the eroding action of violent fall storms which have caused significant damage to the community in previous years. For example, a recent storm in fall 2017 destroyed the road leading from the village to their waste-disposal site, which runs between the coast and their airstrip (Darlene Tocktoo-Turner, pers. comm.).

Light slush ice on the water is very difficult to detect in the low-resolution satellite images. Thus, it was not until 25 November 2006, more than ten days after

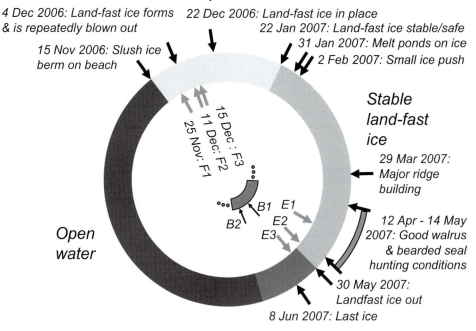

Figure 3.3 Schematic of the seasonal ice cycle at Wales in 2006/07, based on observations by Winton Weyapuk, Jr. (outermost set of black arrows), satellite remote sensing data (inner set of grey arrows), and Alfred Bailey's observations in 1922 (innermost purple circle, thin black arrows correspond to specific observations/photos). Abbreviations: F1 – First ice growing out from shore, F2 – Start of freeze-up offshore, F3 – Ice edge south of Bering Strait; E1 – Ice edge reaches Bering Strait during ice retreat on 7 May 2007, E2 – Ice edge pulls away from Bering Strait to North on 21 May 2007, E3 – Last ice off Wales (8 June 2007); B1 – "Winter conditions still prevailed on June 3, 1922" (Alfred Bailey), B2 – Land-fast ice still in place 16 June 1922 (Alfred Bailey). [A black and white version of this figure will appear in some formats. For the colour version, please refer to the plate section.]

Weyapuk, that the satellite registered the first formation of ice along the shore, with the start of offshore freeze-up reported on 11 December. This illustrates one lesson from local observation: People on the ground detect and monitor the beginning of the freezing period earlier and in more detail than satellites from space.

Drawing on Weyapuk's observations and expert guidance, we have since conducted a larger study of these ice berms and their importance throughout Western Alaska. This work relied on guidance from several Iñupiaq experts

Figure 3.4 A slush berm (*qaimġuq*) forming along the shore on 9 November 2007, with slush ice (*qinu*) and small pieces of pancake ice (*nutiġaġuugvik*) moving with the current just off the slush berm. [A black and white version of this figure will appear in some formats. For the colour version, please refer to the plate section.]
Photo by Winton Weyapuk, Jr

regarding the role such berms play in protecting the shoreline or limiting access to the beach. The long record of Weyapuk's observations at Wales was particularly valuable, as it informed the development of a conceptual model of ice berm formation and allowed comparison with records of weather, ice and ocean conditions (Eerkes-Medrano et al., 2017).

Local experts in Wales indicated repeatedly that fall freeze-up has been delayed in recent years by up to several weeks, compared to past decades. Instrumental observations and satellite images over the past decade also confirm a much later onset of ice formation across the Bering Strait area and northern Alaska (Eicken et al., 2014). Drawing on the observations by Weyapuk and ice experts in other Alaskan communities, we extracted information on the onset and end of the freeze-up and break-up seasons from satellite data. This work shows that for the time period from 1979 through 2013, in northern Alaska, freeze-up occurred later by one to two weeks per decade and break-up occurred earlier by one week per decade (Johnson and Eicken, 2016).

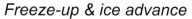

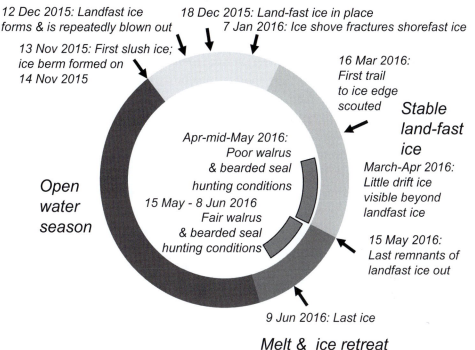

Figure 3.5 Schematic of the seasonal ice cycle at Wales, based on observations by Winton Weyapuk, Jr. in 2015/16. [A black and white version of this figure will appear in some formats. For the colour version, please refer to the plate section.]

While these trends hold across much of Western Alaska and the Arctic, conditions can vary from year to year. For example, comparing the ice seasons in Wales in 2006–2007 and in 2015–2016 (Fig. 3.5) shows that the onset of freeze-up occurred roughly at the same time in mid-November. Break-up, however, occurred earlier by about two weeks in 2016 compared to 2007. The ten years of ice observations contributed by Weyapuk illustrate the pace of change but also the full range of different ice conditions found in any particular year.

Of great importance for a coastal community such as Wales is the formation of shorefast ice (*tuaq*) attached to land and stabilized by the foothold of grounded pressure ridges (*iunit*), formed as a result of offshore pack ice moving in and piling ice into large heaps. In late 2006, it took several weeks from the first appearance of a shorefast ice cover, which subsequently broke off several times, until the shorefast ice remained solidly in place on 22 December. While the ice was thick enough to walk on by 26 December, it was not until almost a full month later, on 22 January 2007, that ice experts in Wales considered the shorefast ice safe as a result of repeated pressure ridge building. These pressure ridges, which stabilize the shorefast ice, also cannot be reliably and easily observed using satellite data. Even so, the lack of massive ridging, possibly a consequence of less thick, old multi-year ice (*utukaq*) making its way down to Bering Strait in recent years, allowed the ice sheet to ride a few feet onshore (*qaupik*) creating a small berm on 2 February. On 4 February 2007, the shorefast ice broke out all the way back to the beach along several miles north of town.

Weyapuk also recorded a warming episode with rain a few days earlier that resulted in ponds forming on the ice surface. While not unheard of in the past, such mid-winter warming events, occasionally accompanied by rain, appear to be more frequent in recent years. Episodes of wet surface snow and ice (*mizagluk*) can pose danger to hunters and may harm the survival of ringed seal pups in their dens. In contrast with the spring melt, which is typically detected well by satellite, those winter warming events may be difficult to detect other than through local, ground-based observations.

In fall and winter of 2015–2016, both the timing of freeze-up and the set-up of the shorefast ice were quite similar to that in 2006–2007 (Figs. 3.3 and 3.5). Unstable shorefast ice with repeated break-out events and an ice shove (*qaupik*) that created openings and detached the shorefast ice near the beach (Fig. 3.6) limited the community's use of the shorefast ice just as in 2006–2007. Later in spring, the lack of shorefast ice stability resulted in ice breaking off at a boat launch (*pituqi*) on the ice at the end of an ice trail put in by hunters from Wales.

In late March 2007, further ridging along the shorefast ice edge helped stabilize the ice, coinciding with the early stages of establishing trails for hauling boats and gear out to a prospective launching site. Hunting with boats launched from the shorefast ice depends on ice and weather conditions suitable for safe launching, passage and access to marine mammals. For approximately one month between mid-April and mid-May 2007, conditions were quite favourable for hunters in Wales. After that date, a combination of adverse weather, offshore ice conditions and the decay of shorefast ice which broke out on 30 May (see Fig. 3.7) made hunting difficult. More importantly, after the removal of the shorefast ice, the presence of drifting ice within the hunting waters near the village was limited to a mere nine days, with the last floating ice seen on 8 June 2007. In 2007, this extremely short transitional period between spring ice conditions and the complete removal of sea ice was considered unusual. Such conditions represent a major disadvantage to subsistence hunters, limiting access to ice-associated marine mammals and eliminating a solid ice platform on which killed animals can be butchered.

The satellite data for 2007 show that the ice edge started to pull away towards the north of Bering Strait around 21 May. The last time the ice was seen off Wales

Figure 3.6 Ice ridge formed as a result of an ice shove event (*qaupik*) on 8 January 2016. Note that large cracks appeared in the shorefast ice as a result of the ice shove, with shorefast ice detaching from shore.
Photo by Winton Weyapuk, Jr

on the satellite images was on 2 June, while observers on the ground reported its presence for almost another week. The weather station at Tin City recorded the start of the melt season as 20 May 2007. Observations of sea ice during the melt season by satellite are difficult because of the presence of water at the surface, masking the signature of ice floes in remote sensing data. Moreover, even the most sophisticated satellites have difficulty distinguishing between the different types of ice that are important to coastal communities and identified in hunter vocabulary. Hence, as demonstrated in a study (Johnson and Eicken, 2016), there is great value in more detailed analysis that combines both satellite data and ground observations by local observers, distinguishing between different types and developmental stages of ice.

The combination of satellite data and local observers' reports also highlight how rapidly the ice edge moved north in spring 2007. In subsequent years, 2017 set a record for the earliest, most extensive spring ice retreat in the region. In Wales and other Alaskan coastal communities, this rapid loss of ice in spring and the absence of lingering ice during summer greatly impacts life and subsistence activities that

Figure 3.7 Shorefast ice (*tuaq*) disintegrates, with the remaining broken pieces (*tuwaiġnit*) drifting away. Scattered pack ice (*tamalaaniqtuaq*) has been driven back close to the shore by northern wind on 7 June 2007, with some floes grounded in shallow water (*ikilituat*). The pack ice that has drifted north and then drifted back south by wind-driven current is called *atitiqtuaq*. [A black and white version of this figure will appear in some formats. For the colour version, please refer to the plate section.] Photo by Winton Weyapuk, Jr

depend on the presence of ice for successful hunting for ice-associated animals, such as the bearded seal and walrus. More importantly, the winters of 2017/18 and 2018/19 were characterized by a near-complete absence of sea ice from the Bering Sea, with dramatic consequences for coastal communities and the marine ecosystem (Cornwall, 2019).

Windows into the Past: Alfred Bailey in Wales, 1922

Working with local partners and learning about Indigenous visions of ice may help scientists advance their knowledge in many areas. One of the most promising applications is to expand the time range of current scientific records. Consistent satellite records of sea ice have only been available since the late 1970s. Reports from local weather stations, with rare exceptions, are silent on the status of ice, its dynamics and major types. Many historical sea ice charts of the early explorers cover the High Arctic areas that are far away from the coastal zone and from the practical issues of today. Thus, applying the knowledge of Arctic residents and scientists interested in former ice conditions in the North may literally open certain 'windows' into the past that are, otherwise, closed.

On 6 April 1922, Alfred M. Bailey, a young biologist from Colorado's Denver Museum of Natural History, arrived in Wales. Bailey (1894–1978), who was then twenty-eight years old, and his Iñupiat companions, reached Wales from the northern Alaskan village of Wainright at the end of a 1200 km (800 mile) dog-sled journey along the frozen Chukchi Sea coast.

Wales was the last stop on Bailey's three-year fieldwork in Alaska that brought him first to south-eastern Alaska in 1919–1920 and, later, in 1921–1922 to Bering Strait, Siberia, Point Barrow (now Utqiaġvik), Wainright and finally, Wales. Bailey was a curator of birds and mammals and the main objective of his Alaskan voyage was to bring back to the Denver Museum skins and skulls of the largest Alaskan mammals – walruses, polar bears, caribou and seals. At the museum, they were used in taxidermied dioramas: exhibit displays of wild animals in their recreated 'live' natural settings.

From early May until late June 1922, Bailey went out hunting with the Kiŋikmiut crews almost every day, until the end of the season. Eventually, Kiŋikmiut hunters were successful in catching enough walruses and seals to support their families for the year and for Bailey to secure the specimens he needed for the future dioramas. During this time, Bailey kept a detailed diary and, in addition, took almost 140 black-and-white photographs of hunters on ice and in boats, people and houses in the village, and the scenery around Wales.

Bailey's Alaskan diaries and photographs are housed at what is now the Denver Museum of Nature and Science (DMNS). Most of Bailey's photographs have been digitized and many images are publicly available on the museum's website at http://dmns.lunaimaging.com/luna/servlet/DMNSDMS including over 120 photos from Wales. Over the years, Bailey used his Alaskan photographs as illustrations to his various publications (Bailey 1933, 1943, 1971), but they remained unknown to the people of Wales until our joint project between 2005 and 2012.

In 2005, Igor Krupnik surveyed Bailey's photographs and diaries at the DMNS archives. The museum generously agreed to share Bailey's photographs from 1922 with the community of Wales for heritage research and education. Prints of Bailey's photographs were sent to the community. Our local partner Herbert Anungazuk (1945–2010) recommended that we engage Winton Weyapuk, Jr. to work with the photographs and to share them with Kiŋikmiut Elders for their insight and personal stories. Altogether, six experts from Wales – Anungazuk, Weyapuk, Pete Sokienna Sereadlook (1930–2017), his wife Lena Sereadlook, Raymond Seetook, Sr. and Faye Ongtowasruk (1928–2015) – contributed their remarks on Bailey's photos. Their comments helped to vastly expand the value of

Bailey's original short captions and contributed additional new knowledge on the environmental and social conditions at Wales in spring 1922.

For our main publication from this co-production effort, the 'Wales Inupiaq Sea Ice Dictionary' (Weyapuk and Krupnik, 2012; see Krupnik, Chapter 5), we selected seventeen photographs taken in 1922 that feature sea ice conditions, subsistence hunting activities and village scenes. All of the photographs were dated, either by the specific date inscribed by Bailey or at least by month. That allowed us to compare them to several dozen contemporary photos taken by Weyapuk during his observations in winter-spring of 2007–2008 and in late 2008. Weyapuk's comments, together with remarks from our senior project participants, such as Pete Sereadlook and Faye Ongtowasruk, provided context for these comparisons and offered insights into much more than ice and weather, as our Kiŋikmiut partners also spoke about how cold weather clothing and hunting equipment had changed.

Though Bailey did not record air temperature, he made numerous references to the wind direction, status of sea ice, ocean currents and overall weather conditions on his numerous hunting trips. These records can often be linked to his photographs taken on the same day. A trained naturalist, Bailey also recorded the timing of the spring arrival of Arctic birds and mammals, as well as the dates of nesting for many bird species. These dates of 1922 can be compared to the present-day timing of similar events. Thanks to this combination of various sources, Bailey's records from Wales offer a rare chance to document the scope of change, at least during the transition from winter to spring to summer exactly one hundred years ago.

Although Bailey's photographs from Wales limit our 'window' into the past ice conditions to merely three months (April to late June) of a single ice season, even as a momentary glimpse, they indicate significant differences between spring 1922 and 2007 or 2016. In spring 2007 and again in 2016, shorefast ice (*tuaq*) had already completely disappeared two to four weeks prior to when Bailey's photo (Fig. 3.8) still shows a solid shorefast ice platform likely to persist for several more weeks. Bailey refers to the ice being in a solid 'winter' state on 3 June 1922, that is, many weeks later than in 2007 and 2016. In 1922, hunting in the ice pack was possible until late June, again, almost a month later in the season compared to recent years, and possibly much longer.

Another facet of the seasonal ice cycle in 1922 can be inferred through Bailey's photos featuring hunters wearing snowshoes (*taglut*) on young ice in the springtime (Weyapuk and Krupnik, 2012: 11, 80). Evidently, the spring melt in those days went through a longer, more drawn-out process of thawing and refreezing that led to the widespread formation of young unstable ice, *siguliaq*, in April and May. Today, spring ice disintegration is so rapid in early-mid May, and even in April, that there is little need for special implements to enable hunters to safely walk on thin, unstable spring ice.

Figure 3.8 Kiŋikmiut hunting crew with their large whaling skin-boat (*umiaq*) at the edge of the shorefast ice on 16 June 1922.
Photo: Alfred Bailey, BA-21-327; Denver Museum of Nature & Science

While spring 1922 may seem, perhaps, anomalously cold, other records from the same era indicate that it reflected a rather 'normal' condition early in the past century. A year prior, Bailey paid a brief visit to Wales on the 'U.S.C.G.C. Bear' and reported heavy drifting ice south of King Island on 27 June 1921 (Bailey 1933, 1971). A few days later, on 30 June 1921, the 'Bear' steered from St. Lawrence Island to Siberia through dense ice floes (Burnham, 1929: 53). At Cape Dezhnev, right across from Wales on the Russian side of Bering Strait, the ice cleared up on 24 June 1921, that is, more or less at the same time or slightly earlier than in Wales in spring 1922.

Several other dates identified in Bailey's photographs came as a surprise to today's Inupiat experts, since in recent years many subsistence activities featured by Bailey take place a few weeks, if not a full month, earlier. To illustrate the scope of change, environmental and cultural alike, we placed Bailey's photos of 1922 beside today's pictures of the more or less similar settings taken by Weyapuk, who also commented on both Bailey's and his own photographs. Most of the photo pairs reveal dramatic differences in ice, snow and weather conditions between 1922 and recent years. A photograph of the village of Wales taken on 1 June 1922 shows all of the houses covered in deep snow, with an unbroken expanse of shorefast ice and no open water on the horizon (Bailey, 1943: 34). Another photo by Bailey dated June 1922 (BA21–619) featured the Wales school building with a group of students and teachers standing on a pile of snow up to the building's roof and the hills in the background completely covered by packed snow. In 2016, the snow in Wales melted away by late May, and in spring 2017 by late April, leaving but small patches on the ground.

Weyapuk's response to another of Bailey's photos featuring Wales' hunters butchering walruses on a large ice floe in June 1922 was quite illuminating. 'This is late June?! Wow! That is pretty late still to have pack ice compared to now. By late June in recent years there's been no ice at all to speak of.'

Weyapuk also noted much colder temperatures in May– June 1922 compared to the present, judged from the status of ice and snow and the clothing that people wore during the spring hunting. 'It [Bailey's caption to the photo – HE-IK] says "early May" – must have been cold at that time. They [the hunters on ice] are dressed pretty warm. The snow on these blocks of ice hasn't even started melting.' On another photograph (BA 21-597) featuring hunters butchering seals on an ice floe, he commented: 'It says "June 1922". Compared to now, that looks like about a month late in conditions and activities. ... The ice surface still looks like it's covered in snow; not in these days.'

Some years prior to or shortly after 1922 turned out to be even colder at Wales and in the Bering Strait region. In spring 1918, the floating ice was gone at Cape Dezhnev by 9 July; and in summer 1926, the shorefast ice was not broken until 23 July, so that the last drifting ice floes disappeared from the area on 6 August 1926. In the past decade, not a single year would have had pack ice in Bering Strait in late June and since 2010, the area has become ice free around mid-June or even earlier. Without today's ground observations that were diligently taken by Weyapuk for ten years (2006–2016), the comparison of these early ice records with modern satellite images and other instrumental data would have been severely limited.

In a similar way, other historical sources may be carefully re-read through Kiŋikmiut eyes to advance our knowledge about past ice conditions. The collection of letters and diaries written by Ellen Kittredge Lopp from Wales between 1892 and 1902 (Lopp and Lopp, 2001) contains numerous references to ice and weather events, ice conditions and the start of spring whaling and walrus hunting in the community. They serve as good evidence of the spring ice dynamics in the Wales area some 120 years ago. For example, the Kiŋikmiut crews were actively hunting walrus on the moving ice floes in early-mid June, even in late June, in 1893, 1898 and 1901, exactly as they did in June 1922, during Bailey's visit.

Today, walrus commonly pass northward through Bering Strait on the retreating ice in mid- or late May and by early June, there is no ice left off Wales anymore.

It would be even more fruitful for today's Indigenous experts to apply their knowledge to interpret the early weather logs of the first missionaries and teachers, as has been done in Gambell, St. Lawrence Island, Alaska (Oozeva et al., 2004). It would extend current understandings of ice dynamics in the Bering Strait area more than 130 years into the past, back to the time of commercial whalers and when teachers first settled in Native communities in the 1890s.

Sea Ice for Walrus Outlook (SIWO)

Since spring 2010, some of the observational entries by Weyapuk and other Native Alaskan village observers for the SIZONet and SIKU projects have been systematically shared on the new community-focused web portal and archive developed for another collaborative initiative called 'Sea Ice for Walrus Outlook' (SIWO – www.arcus.org/siwo Eicken et al., 2011). The 'Sea Ice for Walrus Outlook' (SIWO) was an offshoot of a much larger US-led observational and circumpolar initiative known as the 'Sea Ice Outlook' that was started in 2008 (Calder et al., 2011) and continues to the present.

SIWO was tailored for a particular area: the northernmost portion of the Bering Sea, Bering Strait, and the adjacent southern portion of the Chukchi Sea. It was also designed as a new web-based resource for a particular audience – Alaska Native subsistence hunters, coastal communities and other stakeholders interested in sea ice dynamics, Pacific walrus subsistence hunting and management. It is a web-based network of local community monitors who report their weekly observational entries about sea ice and weather conditions and subsistence hunting in the area during the main spring walrus migration, originally from mid-April until mid-June. Started in spring 2010, officially after the completion of IPY 2007–2008, SIWO continues to this day and has accumulated a remarkable archive of Indigenous sea ice and subsistence observations.

SIWO is a direct descendant of the same IPY projects, SIZONet and SIKU, which pioneered extended collaboration and knowledge co-production with Indigenous experts in Alaska and the Russian Chukotka during the IPY years. It builds upon many years of successful partnership among sea ice and walrus scientists, subsistence users, local Indigenous communities, weather forecasting and game protection agencies, anthropologists and heritage documentation specialists (Metcalf and Krupnik, 2003; Oozeva et al., 2004; Krupnik and Ray, 2007; Eicken et al., 2009, 2011). It was first discussed with representatives from several Bering Strait communities at a meeting supported by the Eskimo Walrus Commission and the US National Science Foundation in Nome, Alaska in January

2010. The template for observations and a new SIWO website were quickly developed and the first weekly sea ice and walrus distribution assessment went online on 2 April 2010 (Eicken et al., 2011).

Originally, Indigenous observers from four Alaskan communities – Wales, Shishmaref, Gambell and Nome – agreed to contribute their weekly summaries of sea ice and walrus hunting conditions. These contributions were integrated with complementary assessments from scientists and observers on ships at sea, at the Alaska National Weather Service (NWS) headquarters in Anchorage and at the University of Alaska Fairbanks, who were using satellite imagery, coastal radars and airborne observations. Weyapuk became the key SIWO contributor for his home community of Wales, together with Paul Apangalook and Merle Apassingok for Gambell, Curtis Nayokpuk for Shishmaref and Fred Tocktoo for Nome.

The SIWO weekly instalments feature improved local weather forecasts and detailed assessments of local sea ice conditions relevant to walrus distribution and migration in the northern Bering Sea and southern Chukchi Sea region adjacent to north-western Alaska and northeastern Russia (Chukchi Peninsula) from April through mid-June. This period was selected to match the interest of Alaska Native hunters who pursue walrus primarily during the peak of the spring migration during break-up and northward retreat of ice in the Bering and Chukchi Seas (Metcalf and Robards, 2008). Each weekly summary on the SIWO webpage (see Fig. 3.9) includes an assessment of the current ice conditions relevant to distribution and access to walrus; a ten-day outlook of wind conditions (speed and direction); and up-to-date satellite imagery for the Bering Strait and St. Lawrence Island, which are two regions of the most interest to coastal Indigenous communities active in the subsistence walrus hunt. Written observations of ice development from Alaska Native hunters, sea ice experts, National Oceanic and Atmospheric Administration (NOAA), National Weather Service (NWS) and university researchers were important elements of the weekly entries.

Though originally designed as a small pilot project, SIWO succeeded in carrying on the legacy partnership and knowledge co-production developed during the IPY 2007–2008 and in terms of making polar research a collaborative enterprise, relevant and valuable to local stakeholders in Indigenous communities. It encouraged data sharing across science disciplines (ocean and ice studies, atmospheric science, marine biology, anthropology and subsistence research) and between scientists and Indigenous organizations, such as the Eskimo Walrus Commission. SIWO was designed with the intention that it may serve as a prototype of a much broader observational service network that would incorporate Indigenous ice and weather observations into the existing agency-supported weather and ice monitoring and forecasting. Such integration could significantly

Sea Ice for Walrus Outlook (SIWO)

Update

Status: The 2020 SIWO season began on Friday, 27 March. Follow us on Facebook at https://www.facebook.com/seaiceforwalrus for more discussion.

To share comments or images about the conditions in your area, send them to Lisa Sheffield Guy (lisa@arcus.org) or join the conversation on the SIWO Facebook page: https://www.facebook.com/seaiceforwalrus.

Subscribe

Subscribe here to receive new SIWO reports via email.

Enter your email address:

Subscribe

Printer Friendly/Low-Bandwidth Version

Overview

The Sea Ice for Walrus Outlook (SIWO) is a resource for Alaska Native subsistence hunters, coastal communities, and others interested in sea ice and walrus. The SIWO provides weekly reports during the spring sea ice season with information on weather and sea ice conditions relevant to walrus in the northern Bering Sea and southern Chukchi Sea regions of Alaska.

The Outlooks are produced with information on weather and sea ice conditions provided by the National Weather Service - Alaska Region and Alaska Native sea ice experts. SIWO is managed by the Arctic Research Consortium of the U.S. (ARCUS), in partnership with the Eskimo Walrus Commission, the National Weather Service, the University of Alaska Fairbanks, and local observers. Funding for SIWO is provided to ARCUS by the National Science Foundation's Division of Arctic Sciences (PLR-1928794).

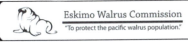

Assessment of Current Ice Conditions Relevant to Distribution and Access of Walrus

Printer Friendly/Low-Bandwidth Version
1 May 2020 - Current Conditions:
Click the name of each community below to view more frequently updated and detailed information from the National Weather Service.

Synopsis High pressure will remain over the northern Bering Sea, while a low moves eastward across the southern Bering Sea through Monday.

Near St. Lawrence Island

Satellite imagery from the evening of 29 April shows the shorefast ice on the north side of the island remains intact. From Gambell to Kangee Camp it extends from about 1.5 miles to 4 miles off-shore. Between Kangee Camp and Savoonga, there is an area of compact pack ice with small to medium floes. Shorefast ice continues from Savoonga past Camp Kulowiye, extending southward across the eastern edge of the island and extends roughly 1 to 3 miles away from the coast except for off-shore Camp Iveetok, where it extends 6 to 7 miles offshore. There is also shorefast ice on the south side of the island, extending 2 to 3 miles from shore from Powooiliak Camp and east. South of the shorefast ice in this area, exists an area of open pack ice, mainly new. This open to very open pack ice varies in distance, 15 to 25 miles away from the coast. Close pack ice is 6 to 10 miles away from the shorefast of Siknik camp. Otherwise, consolidated to compact pack ice consisting of big to giant floes surrounds St. Lawrence Island.

Figure 3.9 Screenshot of SIWO website.

augment and improve the design and implementation of an Arctic observing system from broad to local spatial and temporal scales (Eicken, 2013).

SIWO's continuous operation for eleven years since 2010 is a tribute to the dedication of local experts such as Winton Weyapuk, Jr., who was one of the most regular and prolific SIWO contributors until his last days. Another tribute to Weyapuk's lasting legacy was the decision of a younger hunter from Wales, Robert Tokeinna, Jr., to step into Weyapuk's shoes as the SIWO observer for Wales in spring 2017. It is our shared hope that the SIWO initiative continues and that the contributions from local observers will build on the legacy of the partnership started by Weyapuk. The SIWO may thus become a lasting fixture of the polar 'services' provided by the federal and state agencies to local communities for years and decades ahead.

Conclusions

Local sea ice knowledge, such as observations, terms for many types of ice in Indigenous languages and stories told by experienced hunters and Elders, remains critical to the use of sea ice in northern communities, as well as to scientists studying Arctic change. Local experts use Indigenous terms for sea ice as a tribute to their Elders and as praise for the youth, who continue hunting in today's less stable and more dangerous sea ice (Weyapuk and Krupnik, 2012). Their willingness to work with scientists should be viewed in this light, as an opportunity to collaborate and to increase the reciprocal value of sharing across different knowledge systems. This is what we have learned over years of working with Winton Weyapuk, Jr., a dedicated expert, keen naturalist and supporter of his language and cultural tradition.

The co-production of knowledge also reveals challenges and opportunities that go hand-in-hand with the rapid change that is sweeping across the North and that require careful and systematic action. The earlier and more pronounced expression of global climate trends in the Arctic has already forced the people of northern Alaska to devise ways in which traditional lifestyles can be adapted in the face of change. They have demonstrated great resources of innovation and resilience in the use of sea ice. This stock of resilience and innovation possessed by polar residents will be, perhaps, their strongest asset in facing the challenges of future change.

Such a perspective is not meant to trivialize many prospective impacts of rapid climate or social change across the polar regions and the upheaval it may bring to people's lives. Other challenges include relocation of some coastal communities under the threat of coastal erosion, potential shifts in distribution of marine mammals owing to new ice patterns, or the socio-economic transformations brought about by environmental change; indicating that a concerted and collaborative effort by many interested parties is needed to deal with adversity in the face of change.

We see our decade-long partnership with Winton Weyapuk, Jr. and the Native Community of Wales as a contribution to this common effort. It points to the importance of Indigenous and local knowledge and of a long history of observing and responding to change that may be crucial to people's preparedness for what will likely be a vastly different Arctic that their children and grandchildren are to inherit. It also illustrates that our understanding of the many facets of this changing system would have hardly advanced beyond the level of large-scale but low-resolution general models if not for the input of Weyapuk and other Indigenous partners to the shared co-production of knowledge in the process of collaborative research.

Acknowledgements

We appreciate and recognise the important role that Indigenous experts from Wales have played in the success of this collaboration. These include, besides Weyapuk, Jr., Peter Sereadlook, Raymond Seetok, Eugene Angnabooguk, Herbert Anungazuk, Faye Ongtowasruk and several others. Davis Ongtowasruk and Michael Ahkinga were helpful in our fieldwork on sea ice. We appreciate the support of the Native Village of Wales and the Bering Straits School District. We are grateful for funding support from the National Science Foundation's Arctic Observing Network Program, the University of Alaska's IPY Initiative and the Smithsonian Institution. The Denver Museum of Nature and Science generously shared photos from the Alfred M. Bailey collection for our work in Wales and for future publications, and Chris Petrich offered his photograph of Winton Weyapuk to illustrate this paper.

References

Bailey, A. M. 1933. A cruise of the "Bear". *Natural History*, 3(5): 497–510.
Bailey, A. M. 1943. The birds of Cape Prince of Wales, Alaska. *Proceedings of the Colorado Museum of Natural History*, 18(1).
Bailey, A. M. 1971. Field work of a museum naturalist. *Museum Pictorial* 22. Denver, CO: Denver Museum of Natural History.
Bogoslovskaya, L. S. and Krupnik, I. (eds.) 2013. *Nashi L'dy, Snega i Vetry [Our Ice, Snow and Winds.] Indigenous and Academic Knowledge on Ice-Scapes and Climate of Eastern Chukotka.* Moscow: Russian Heritage Institute.
Burnham, J. B. 1929. *The Rim of Mystery: A Hunter's Wandering in Unknown Siberian Asia.* New York: G. P. Putnam's Sons.
Calder, J., Eicken, H. and Overland, J. 2011. The sea ice outlook. In Krupnik, I., Allison, I., Bell, R., Cutler, P., Hik, D., López-Martínez, J., Rachold, V., Sarukhanian, E. and

Summerhayes, C. (eds.) *Understanding Earth's Polar Challenges: International Polar Year 2007–2008*. Rovaniemi: University of the Arctic and Edmonton: CCI Press, pp. 405–410.

Comiso, J. C. 2017. *Bootstrap Sea Ice Concentrations from Nimbus-7 SMMR and DMSP SSM/I-SSMIS, Version 3. 1978–1979*. Boulder, CO: NASA National Snow and Ice Data Center Distributed Active Archive Center. http://dx.doi.org/10.5067/7Q8HCCWS4I0R

Cornwall, W. 2019. Vanishing Bering Sea ice poses climate puzzle. *Science*, 364(6441): 616–617. http://DOI.org/10.1126/science.364.6441.616

Eerkes-Medrano, L., Atkinson, D. E., Eicken, H., Nayokpuk, B., Sookiayak, H., Ungott, E. and Weyapuk, Jr., W. 2017. Slush-ice berm formation on the west coast of Alaska. *Arctic*, 70(2): 190–202. https://doi.org/10.14430/arctic4644

Eicken, H. 2010. Indigenous knowledge and sea ice science: What can we learn from Indigenous ice users? In Krupnik, I., Aporta, C., Gearheard, S., Laidler, G. J. and Kielsen Holm, L. (eds.) *SIKU: Knowing Our Ice. Documenting Inuit Sea Ice Knowledge and Use*. Dordrecht: Springer, pp. 357–376.

Eicken, H. 2013. Arctic sea ice needs better forecasts. *Nature*, 497: 431–433. https://doi.org/10.1038/497431a

Eicken, H., Lovecraft, A. L. and Druckenmiller, M. L. 2009. Sea-ice system services: A framework to help identify and meet information needs relevant for Arctic observing networks. *Arctic*, 62(2): 119–136. https://doi.org/10.14430/arctic126

Eicken, H., Hufford, G., Metcalf, V., Moore, S., Overland, J. and Wiggins, H. 2011. Sea Ice for Walrus Outlook (SIWO). In Krupnik, I., Allison, I., Bell, R., Cutler, P., Hik, D., López-Martínez, J., Rachold, V., Sarukhanian, E. and Summerhayes, C. (eds.) *Understanding Earth's Polar Challenges: International Polar Year 2007–2008*. Rovaniemi: University of the Arctic and Edmonton: CCI Press, pp. 405–410.

Eicken, H., Kaufman, M., Krupnik, I., Pulsifer, P., Apangalook, L., Apangalook, P., Weyapuk, Jr., W. and Leavitt, J. 2014. A framework and database for community sea ice observations in a changing Arctic: An Alaskan prototype for multiple users. *Polar Geography*, 37(1): 5–27. https://doi.org/10.1080/1088937X.2013.873090

Fox, S. 2002. These are things that are really happening: Inuit perspectives on the evidence and impacts of climate change in Nunavut. In Krupnik, I. and Jolly, D. (eds.) *The Earth Is Faster Now: Indigenous Observations of Arctic Environmental Change*. Fairbanks, AK: ARCUS, pp. 12–53.

Gearheard (Fox), S., Holm, L. K., Huntington, H. P., Leavitt, J. M. and Mahoney, A. R. (eds.) 2013. *The Meaning of Ice. People and Sea Ice in Three Arctic Communities*. Hanover, NH: International Polar Institute.

Gearheard (Fox), S., Matumeak, W., Angutikjuaq, I., Maslanik, J., Huntington, H. P., Leavitt, J., Kagak, D. M., Tigullaraq, G. and Barry, R. G. 2006. "It's not that simple": A collaborative comparison of sea ice environments, their uses, observed changes, and adaptations in Barrow, Alaska, USA, and Clyde River, Nunavut, Canada. *Ambio*, 35: 203–211. https://doi.org/10.1579/0044-7447(2006)35[203:INTSAC]2.0.CO;2

George, C. J., Huntington, H. P., Brewster, K., Eicken, H., Norton, D. W. and Glenn, R. 2004. Observations on shorefast ice dynamics in Arctic Alaska and the responses of the Iñupiat hunting community. *Arctic*, 57(4): 363–374. https://doi.org/10.14430/arctic514

Johnson, M. and Eicken, H. 2016. Estimating Arctic sea-ice freeze-up and break-up from the satellite record: A comparison of different approaches in the Chukchi and Beaufort Seas. *Elementa*, 4: 124. http://doi.org/10.12952/journal.elementa.000124.

Krupnik, I. 2000. Native perspectives on the climate and sea-ice change. In H. P. Huntington (ed.) *Impacts of Changes in Sea Ice and Other Environmental*

Parameters in the Arctic. Background report to the international Arctic sea-ice change workshop. Washington, DC: Marine Mammal Commission, pp. 6–14.

Krupnik, I. 2002. Watching ice and weather our way: Some lessons from Yupik observations of sea ice and weather on St. Lawrence Island, Alaska. In Krupnik, I. and Jolly, D. (eds.) *The Earth Is Faster Now: Indigenous Observations of Arctic Environmental Change.* Fairbanks, AK: ARCUS, pp. 156–199

Krupnik, I. 2009. "The way we see it coming": Building the legacy of Indigenous observations in IPY 2007-2008. In Krupnik, I., Lang, M. and Miller, S. (eds.) *Smithsonian at the Poles: Contributions to International Polar Year Science.* Smithsonian Institution Scholarly Press, pp. 129–142.

Krupnik, I. 2017. Winton Weyapuk, Jr., 1950–2016: Naturalist, observer and community pillar. *Arctic Studies Center Newsletter*, 24: 66–67.

Krupnik, I., and Jolly, D. (eds.) 2002. *The Earth is Faster Now: Indigenous Observations of Arctic Environmental Change.* Fairbanks, AK: Arctic Research Consortium of the United States.

Krupnik I. and Ray, G. C. 2007. Pacific walrus, Indigenous hunters, and climate change: Bridging scientific and Indigenous knowledge. *Deep-Sea Research, Part II*, 54: 2946–2957. https://doi.org/10.1016/j.dsr2.2007.08.011

Krupnik, I., Aporta, C., Gearheard, S., Laidler, G. and Holm, L. K. (eds.) 2010a. *SIKU: Knowing Our Ice. Documenting Inuit Sea Ice Knowledge and Use.* Dordrecht: Springer.

Krupnik, I., Apangalook, L., and Apangalook, P. 2010b. "It's cold, but not cold enough": Observing ice and climate change in Gambell, Alaska, in IPY 2007–2008 and beyond. In Krupnik, I., Aporta, C., Gearheard, S., Laidler, G. and Holm, L. K. (eds.) *SIKU: Knowing Our Ice. Documenting Inuit Sea Ice Knowledge and Use.* Dordrecht: Springer Netherlands, pp. 81–114. https://doi.org/10.1007/978-90-481-8587-0_4

Laidler, G. J. 2006. Inuit and scientific perspectives on the relationships between sea ice and climate change: An ideal complement? *Climatic Change*, 78: 407–444. https://doi.org/10.1007/s10584–006-9064-z

Laidler, G. J. and Elee, P. 2008. Human geographies of sea ice: Freeze/thaw processes around Cape Dorset, Nunavut, Canada. *Polar Record*, 44(228): 51–76. https://doi.org/10.1017/S0032247407007061

Laidler, G. J. and Ikummaq, T. 2008. Human geographies of sea ice: Freeze/thaw processes around Igloolik, Nunavut, Canada. *Polar Record*, 44(229): 127–153. https://doi.org/10.1017/S0032247407007152

Lopp Kittredge, E. L. and Lopp, W. T. 2001. *Ice Window: Letters from a Bering Strait Village.* Fairbanks: University of Alaska Press.

McDonald, M., Arragutainaq, L. and Novalinga, Z. (compilers). 1997. *Voices from the Bay: Traditional Ecological Knowledge of Inuit and Cree in the Hudson Bay Bioregion.* Ottawa, Ontario: Canadian Arctic Resources Committee.

Metcalf, V. and Krupnik, I. (eds.) 2003. *Pacific Walrus: Conserving Our Culture through Traditional Management.* Report produced by Eskimo Walrus Commission, Kawerak, Inc. under the grant from the U. S. Fish and Wildlife Service, Section 119, Cooperative Agreement #701813J506. Alaska: Eskimo Walrus Commission.

Metcalf, V. and Robards, M. 2008. Sustaining a healthy human-walrus relationship in a dynamic environment: Challenges for co-management. *Ecological Applications*, 18(2): 148–156. https://doi.org/10.1890/06-0642.1

National Snow and Ice Data Center. 2019. *Arctic Sea Ice News and Analysis.* Online resource. https://nsidc.org/arcticseaicenews/2019/09/arctic-sea-ice-reaches-second-lowest-minimum-in-satellite-record/ (retrieved 18 April 2020).

Nelson, R. K. 1969. *Hunters of the Northern Ice*. Chicago: University of Chicago Press.

Oozeva, C., Noongwook, C., Noongwook, G., Alowa, C. and Krupnik, I. 2004. *Watching Ice and Weather Our Way/Sikumengllu Eslamengllu Esghapalleghput*. Washington, DC: Arctic Studies Center.

PIOMAS. 2020. Pan-Arctic Ice Ocean Modeling and Assimilation System (PIOMAS) *Arctic sea ice volume reanalysis*, http://psc.apl.uw.edu/wordpress/wp-content/uploads/schweiger/ice_volume/BPIOMASIceVolumeAprSepCurrent.png. (accessed on April 18, 2020)

Stroeve, J., Serreze, M., Drobot, S., Gearheard, S., Holland, M., Maslanik, J., Meier, W. and Scambos, T. 2008. Arctic sea ice extent plummets in 2007. *Eos*, 89(2): 13–20. https://doi.org/10.1029/2008EO020001

Taylor, P. C., Maslowski, W., Perlwitz, J. and Wuebbles, D. J. 2017. Arctic changes and their effects on Alaska and the rest of the United States. In *Climate Science Special Report: Fourth National Climate Assessment* Vol. 1. Washington, DC: U.S. Global Change Research Program, pp. 303–332, doi: 10.7930/J00863GK

Weyapuk, Jr., W. and Krupnik, I. (compilers). 2012. *Kingikmi Sigum Qanuq Ilitaavut: Wales Inupiaq Sea Ice Dictionary*. Washington, DC: Arctic Studies Center, Smithsonian Institution; available at https://jukebox.uaf.edu/site7/sites/default/files/documents/Preserving-our-Knowledge–Wales-Dictionary.pdf

Worby, A. P. and Eicken, H. 2009. Ship-based ice observation programs. In Eicken, H., Gradinger, R., Salganek, M., Shirasawa, K., Perovich, D. and Leppäranta, M. (eds.) *Field Techniques for Sea-Ice Research*. Fairbanks: University of Alaska Press, pp. 365–381.

4

Shaping the Long View: Iñupiat Experts and Scientists Share Ocean Knowledge on Alaska's North Slope

MATTHEW L. DRUCKENMILLER

> Our land was already a very special place, long before the newcomers began showing intense interest upon our land. It is a place that has provided for the spiritual and physical well-being of the people who learned to rely on its renewable resources. We too, like others in far-off lands, taught others who came to us, to survive as we have survived, because there was 'no other way'. Everything that the people ever needed was from the land and the sea; our ways are natural.
>
> Herbert Anungazuk, Iñupiat whaler from Kiŋigin (Wales, Alaska).
> Excerpt from 'Whaling: Indigenous Ways
> to the Present' (Anungazuk, 2003).

Introduction

The Iñupiat of Alaska live within a land of extremes. Extreme cold, wind and distances shape the roles and values of their communities. Iñupiat, a derivation of Inuit, refers to the Indigenous peoples of North Alaska that inhabit the Arctic and subarctic regions extending from the coasts of Norton Sound within the Bering Sea to Alaska's northernmost border with Canada. The vast majority of Iñupiat communities live at the edge of the tundra, along an ocean that is transformed by sea ice for most of the year. The ocean provides much of their nourishment, mobility and freedom in the far North. Frozen seascapes represent home and define ancestral and spiritual connections (see Gearheard et al., 2013). Yet, to be on sea ice is to be in the realm of the ocean. When venturing onto ice, a new sense of awareness is required and the rules for assessing risk change drastically. For Iñupiat hunters of seal, walrus, whale and fish, the ice is an environment they must experience and learn to use for survival. The Iñupiat have a long tradition of sharing knowledge about their environment and the role of people within it. Sharing underlies their cultural values and is central to understanding animal behaviour and migrations, hunting strategies, individual safety and survival in the harshest of conditions. Sharing has also become a way to engage beyond their communities, for which this chapter is evidence.

The Arctic sea ice environment has long been a focus of Western research as a context for documenting and understanding Indigenous knowledge. From the early years of Franz Boas (1888), a rich collection of anthropological studies has focused on individual and community interaction with sea ice for survival and culture, addressing safe methods of ice travel, ice and weather forecasting, marine mammal hunting, and methods of dress and improvisation (e.g., Stefansson, 1919; Foote, 1960; Nelson, 1969; Lowenstein, 1980; Krupnik and Jolly, 2002/2010; Krupnik et al., 2010). As the Arctic ice rapidly recedes and thins (Stroeve et al., 2012; Lindsay and Schweiger, 2015), sea ice has now become an indicator of global climate change and a frontier for new research and activities.

Ongoing changes in the Arctic environment have commanded the attention of global audiences. For the Iñupiat, however, these dramatic changes follow a series of many other types of rapid change encountered within recent generations. The cultural and environmental impacts of commercial whaling, which spanned the period from 1848 to 1910 (see Bockstoce, 1986), the establishment of Western schools, the transition to mixed-economies and the discovery of the Prudhoe Bay Oil Field in 1968 all initiated periods of community restructuring. These times brought changes in infrastructure and socio-political life to the North Slope as well as hardships and traumatic experiences that deeply impacted livelihoods for many generations. Despite a complicated history, the spirit of the whale continues to remain central to Iñupiat well-being, community life, families and spirituality, as well as to their relationship with the landscapes and seascapes that sustain them.

The North Slope coastline is a social-cultural transition zone between sea and land, marked by the location of past and current settlements, burial sites, family hunting locations, traditional places of refuge and places immortalized through traditional stories. The coastline is where the Iñupiat observe, enter and exit the ocean, and hence is interwoven throughout their local and traditional knowledge of the marine environment. Rarely do Iñupiat hunters speak of ocean or ice features, or of a hunting story, without referring to a place along the coast.

North Slope communities and their coastlines are also staging areas for scientists – oceanographers, ice physicists and biologists – who have adopted the Chukchi and Beaufort Seas as their natural laboratories. Yet, the scientists often do not need to venture far to explore interesting questions; it is often the interaction between land, shallow waters, currents, ice and animals that provides researchers with observations needed to better understand the interplay of natural phenomena.

Coastlines and coastal waters thus provide both physical and conceptual spaces to begin and advance knowledge sharing across Indigenous knowledge and science.

This chapter will explore a cross-section of the North Slope's rich history of scientists working with Indigenous Elders and peoples on issues related to local land and seascapes, with specific attention to coastal emergency preparedness. There has been notable success in collaboratively co-assembling different forms of knowledge. The degree to which knowledge co-production has taken place is a much more complicated question. However, I aim to reveal a local history that provides a basis for moving in this direction, which, in my view, requires a clear imperative. On Alaska's North Slope, this urgency is spurred by the prospect of increasing industrial activities in the Arctic, including shipping, tourism, and offshore oil and gas development. Further urgency arises when confronted with the reality that Iñupiat knowledge today is transmitted to younger generations very differently than in previous generations.

A Well-Told History of Working Together (1977 to Present Day)

The North Slope Borough (NSB), based in Utqiaġvik,[1] Alaska (see Fig. 4.1), is the municipal government that encompasses approximately the northern third of Alaska, an area of nearly 24 million hectares. Today, the NSB is one of only a few boroughs or county-level governments (i.e., the largest administrative divisions of a state) in the United States that has a wildlife department. The primary role of the

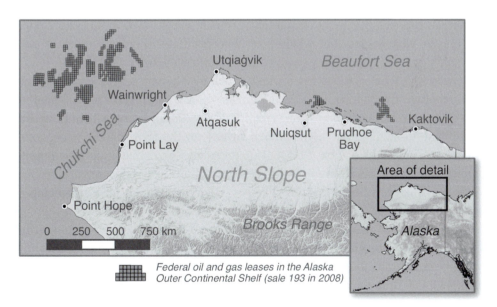

Figure 4.1 Map of Alaska North Slope communities.

NSB's Department of Wildlife Management (DWM) is to monitor populations of fish and wildlife species in collaboration with local residents and the broader scientific community, and to ensure sustainable and continued subsistence harvests. It is with these goals in mind that the NSB engages on issues related to expanding industrial activity and changing land and marine environments resulting from climate warming trends. The local and Indigenous knowledge of the Iñupiat continues to play a guiding role in how the DWM views the best available knowledge to deal with these issues. In fact, the application of local and Indigenous knowledge together with Western science for understanding wildlife was the primary basis for the Department's creation in 1982.

Bowhead whales (*Balaena mysticetus*) – baleen whales that can reach up to 20 m in length – were once abundant throughout Arctic and subarctic seas. Then, between 1848 and 1910, commercial whaling in pursuit of whale oil, and later for baleen used to make dress corsets, decimated the populations (Bockstoce, 1986). The Western Arctic bowhead population, also known as the Bering-Chukchi-Beaufort (BCB) stock in accordance with the seas that provide for its annual migration range, was reduced from approximately 25,000 animals to around only a few thousand (Bockstoce, 1986; Woodby and Botkin, 1993). Following the end of commercial whaling, the Iñupiat continued their traditional subsistence harvesting of the BCB stock. Nearly sixty years later, in 1976 and 1977, scientists from the US National Oceanic and Atmospheric Administration (NOAA) conducted whale censuses using visual sightings from the ice edge and, to a lesser extent, through aerial surveys, and estimated that the whale population was at a critical low number of between 600 to 2,000 animals (Tillman, 1980). Around this same time, the Iñupiat were revitalizing whaling by incorporating different equipment and technologies using money they received from North Slope oil production. In 1977, there were 111 whale strikes (i.e., a whale struck by a hunter, whether harvested or not) reported by Iñupiat whalers (Tillman, 1980). This quickly led to the International Whaling Commission (IWC) limiting Alaska Native whaling for the 1978 hunt through a quota of twenty strikes or fourteen landed whales, or whichever came first (Albert, 2000). This threatened an essential part of Iñupiat culture and their legal ability to fulfil subsistence and livelihood needs.

Iñupiat whalers believed that the scientists' counting methods were significantly underestimating the whale population by not accounting for whales travelling beneath sea ice or within the pack ice, far beyond the coastal lead system. The Iñupiat knew that bowheads were not restricted to open water and could break sea ice with their heads to create small cracks through which to breathe. The lack of data and the differences between how the hunters and scientists understood bowhead behaviour, together with the need to interact with regulatory agencies, led the hunters to create the Alaska Eskimo Whaling Commission (AEWC) (Albert,

2000). The struggle to form the AEWC came during a period of much determined effort, backed by growing political influence by impassioned Iñupiat leaders who had already been fighting to prevent nuclear tests on Native lands and to resolve land tenure disputes with the State of Alaska (Stern, 2004, p. 77). AEWC, through a cooperative agreement with NOAA, obtained authority to regulate the hunt and were instrumental in transferring responsibility for estimating the whale population to the North Slope Borough, which, as a result, created the Borough's DWM. DWM scientists faced logistical and scientific challenges to improving ice-based whale counting methodologies but also were confronted by the need to account for the expert knowledge of local hunters.

Throughout the early 1980s, DWM scientists, guided by community Elders, consulted local hunters and developed observational approaches to assess the validity of what the hunters knew from their generational knowledge, spiritual connections and direct personal experiences observing whales. To their benefit, there were already close working relationships and trust between scientists and local hunters. Scientists from the Naval Arctic Research Laboratory (NARL), which operated in Utqiaġvik between 1947 and 1981, had previously collaborated with the AEWC, the Barrow Whaling Captains Association (BWCA) and Iñupiat hunters to access and perform analysis on bowhead whale tissue samples. Key hunters, such as Harry Brower, Sr., with an interest in working with scientists, facilitated a broader collaboration between Utqiaġvik's whalers and DWM biologists (Brewster, 1998; Albert, 2000). Brower understood that the hunters' observations of whale behaviour, if shared, would be critical to the scientists' understanding of whales (Albert, 2000). Central to the ability of the census to account for whales travelling beneath ice, out of sight of visual observers, was the use of passive acoustic recorders, which had been approved for use by hunters before the 1982 census. Hunters also assisted in locating the best ice edge locations for spotting migrating whales near Utqiaġvik.

Through refined observational techniques that drew upon Indigenous knowledge and experience, combined with powerful statistics, the DWM and its collaborating scientists were able to better estimate the population of bowhead whales during the following decades. In 2001, 10,470 animals were reported (George et al., 2004b) and 16,892 in 2011 (Givens et al., 2013). Not only was the whale population well above previously reported numbers but also the population was steadily increasing. This increase is attributed to the long-term absence of commercial whaling, a relatively pristine environment, a low population relative to carrying capacity and sustainable Indigenous harvesting practices (George et al., 2004b). Over this time, the strike quota for Alaska Native whaling had accordingly increased to levels that more reasonably permitted Iñupiat whalers to fill their ice cellars with harvests to share throughout the community. In 2016, for example, the

IWC set the quota at seventy-five strikes to be distributed across the eleven Alaska Native whaling communities (NOAA, 2016).

This story has been told many times (e.g., Brewster, 1997; Albert, 2000; Huntington, 2013), and serves as perhaps one of the most familiar examples of Alaska Native peoples contributing their knowledge to science to influence scientific understanding, research methodologies and policy. Moreover, this example was instrumental in introducing the concept of 'traditional knowledge' to wider scientific and decision-making audiences in Alaska and beyond.

The collaboration between hunters and biologists on Alaska's North Slope has continued with more recent examples of local and traditional knowledge influencing the scope of science. For example, in contrast with the established knowledge of most marine scientists, Iñupiat hunters have long claimed that bowhead whales have a functioning sense of smell – an understanding gained through careful observation of how whales respond to hunters. Recognizing the implications for how whales locate prey and ultimately how they may respond to industrial activity, NSB scientists and collaborators took a closer look and concluded in 2010 that bowheads, in fact, do have a small but complex olfactory bulb within their brain that allows them to smell (Thewissen et al., 2011). Again, the expert knowledge of Utqiaġvik whalers was largely credited for instigating the scientific investigations behind this finding.

With the prospect of offshore oil and gas development gaining increased attention in preparation for the 2008 Chukchi Sea Outer Continental Shelf Oil and Gas lease sale (Sale 193; see Fig. 4.1), the Alaska Department of Fish and Game, with funding from the US Minerals Management Service (now Bureau of Ocean Energy Management, BOEM), began work with the NSB to study the movement and behaviour of whales using satellite telemetry (Quakenbush et al., 2012). This project included placing satellite transmitters on bowhead whales to send location data via satellites as the whales migrate and pass through areas of proposed development and oil and gas exploration activities. The placement of tags required the navigation of boats safely alongside whales in open water where one could accurately throw a harpoon to attach the transmitter. Owing to the difficulty of the task, Alaskan and Canadian Native hunters (e.g., from Utqiaġvik, Kaktovik and Tuktoyaktuk) played key roles in placing most of the tags and in the overall strategic planning to safely run such a programme. This work has continued through to the present day and has offered an opportunity for scientists to share data with Native whaling communities in near-real-time, resulting in ongoing discussions and email correspondence about the timing and extent of the whales' range, their proximity to villages and the influences of sea ice and other changing conditions on their migration. The collaboration provides hunters new knowledge about the bowhead whale, a glimpse into how scientists work and an opportunity to engage in activities ultimately focused on understanding how to mitigate the adverse impacts of the region's longer term economic prospects on whales, livelihoods and the well-being of their communities.

Institutional Frameworks: Incorporating Whalers' Knowledge

There is little doubt that personal relationships between individual, local Indigenous experts and scientists play an important role in knowledge sharing and co-production. This refers to the intimate side of knowledge co-production where personalities and common everyday gestures of mutual respect and friendship are vital. Close working relationships naturally lead to sharing that exposes each other to elements of their respective knowledge systems, often within the context of jointly observing the environment, discussing wildlife behaviour, recounting traditional stories, or preparing equipment for hunting or travel. However, the manner in which a people collectively develop an identity linked to their knowledge system and, in turn, communicate that identity to the outside world also play roles in positioning their knowledge with regard to issues they deem important to their culture, environment and resources.

Iñupiat representatives from the AEWC, BWCA and NSB play important roles in engaging with regulatory agencies, industry representatives and academia, and in promoting the value of Iñupiat traditional knowledge towards the stewardship and understanding of their land, ocean and living resources. In particular, through these organizations, Iñupiat leaders have established and communicated well-defined priorities that have enabled North Slope communities to assert the value of traditional knowledge to the scientific community and, as a result, have shaped the front line of expanding the way in which Indigenous knowledge is incorporated and respected. For example, any scientific research being conducted on or from the sea ice near Utqiaġvik in springtime requires approval from the BWCA. This process informs the hunters of ongoing and potentially relevant science; provides a mechanism for engagement; and minimizes the likelihood of conflict between research and whaling. Craig George, a long-serving biologist with the NSB Department of Wildlife Management since the late 1970s and one of the most important and persistent links between the local community and scientists, has equated the process of seeking research approval from the BWCA to meeting a 'higher power'. At least, this is what I was warned prior to presenting the proposed topic of my doctoral sea ice studies to the BWCA in 2007. The experience was incredibly humbling as I found myself, a young graduate student, in a room with thirty to fifty Iñupiat whaling captains, expected to justify my research to a community that I was not yet familiar with. However, humility is a necessary precursor for knowledge co-production and for stepping outside the bounds of one's worldview to consider the value of another. George's warning, while in part hyperbole, emphasized the unique nature of the gathering – a place where any ingrained illusions about the authority of Western science and the power of academic institutions rightfully vanish, and the worldview and livelihood priorities of the Iñupiat take full control. The abilities of a scientist to listen and remain flexible are revealed as foundational tenets for crafting partnerships.

The Iñupiat relationship with the whale plays a key role in how they organize and express community and knowledge. Also, it is most often through their whaling identity that Elders emerge as authorities in traditional knowledge and choose to engage with the scientific community. Still, personal histories are also very important. For example, the late Kenneth Toovak (1923–2009) had a long and remarkably diverse career. For many years, he served as a technician on various engineering projects, including with the US Air Force, to establish the North Slope's Distant Early Warning (DEW) system to detect Soviet missiles during the Cold War and with the US Navy to conduct petroleum exploration. Throughout his many years working with scientists, he shared his local knowledge of the Arctic environment with various researchers and, in particular, became a widely recognized expert on sea ice dynamics. For this immense contribution, the University of Alaska Fairbanks granted him an honorary doctorate in 2003, an honour that he shares with others, including Harry Brower, Sr. Such gestures are meaningful in recognizing bridges across knowledge systems. In Alaska and beyond, the knowledge system of North Slope whalers has become a recognized and respected institution – a tradition of sharing, collaboration and recognized authority on the coastal and marine environment of the North Slope – led by adaptive and diplomatic Iñupiat Elders.

The Intersection of Science and Survival: Collaboration on Ice Safety and Hazards

Collaboration between local Indigenous experts and sea ice scientists also has a long history in Utqiaġvik. In the late 1970s, Lou Shapiro and Ron Metzner, scientists studying ice mechanics, worked with Kenneth Toovak to conduct a range of oral history interviews with local hunters on the topic of ice shove events: when wind, the ocean or ice push sea ice onto land, presenting a hazard to infrastructure and human life.[2] In the 1970s and 1980s, Shapiro worked in Utqiaġvik to investigate shorefast ice dynamics using coastal radar. Shorefast ice is the relatively immobile sea ice that develops in waters along the coast by anchoring to the sea floor. It provides an important platform for local travel and hunting, including for bowhead whaling during spring (see Fig. 4.2). The relevance of geophysical approaches to observing and understanding shorefast ice stability and

(a)

(b)

Figure 4.2 (a) An Utqiaġvik whaling crew moving their hunting equipment out onto the shorefast ice in spring; (b) Utqiaġvik, Alaska, a village of approximately 4,000 people, is shown in springtime with a shelf of shorefast ice along the coast. [A black and white version of this figure will appear in some formats. For the colour version, please refer to the plate section.]
Photos by Matthew L. Druckenmiller

associated ice hazards captured the attention of local hunters and of NSB biologists, who, like hunters, spent long periods of time on shorefast ice, exposing themselves to the risk of ice detaching and drifting seaward into remote and dangerous waters (George et al., 2004a).

In 2000, the Barrow Symposium on Sea Ice was organized as a three-day gathering of over thirty Iñupiat ice experts, NSB scientists and sea ice geophysicists to discuss ice hazards in the context of past notable events and to explore areas for potential collaborative sea ice research (Huntington et al., 2001). Around this same time, Hajo Eicken, a sea ice scientist from the University of Alaska Fairbanks (UAF), and Andy Mahoney, then a graduate student working with Eicken, began annual fieldwork in Utqiaġvik to monitor the growth and seasonal evolution of shorefast ice. Their work initiated the Barrow Sea Ice Observatory (Druckenmiller et al., 2009), which has since maintained a long-term record of local sea ice conditions from 1999 to the present day. With the urging of NSB biologist Craig George, Eicken took a strong interest in working more closely with the Utqiaġvik community on issues of local importance. Thus, shorefast ice stability and dynamics became a focus for the sea ice observatory and ultimately the primary topic of Mahoney's doctoral research (see Mahoney, 2006; Fig. 4.3).

Figure 4.3 Ice geophysicist Andy Mahoney and Iñupiat hunter and guide Quuniq Donovan on the shorefast ice off Utqiaġvik.
Photo by Matthew L. Druckenmiller

It was also from the Barrow Symposium on Sea Ice that the suggestion arose for scientists to pay closer attention to the observations and ice use by whalers during their spring bowhead whale hunt. At Utqiaġvik, as in many other Alaska Native whaling communities, hunters pursue migrating whales in *umiaks* (small wooden-framed seal skin boats) launched from camps along the edge of the shorefast ice. The hunters access the ice edge by constructing trails across the ice, often through extensive effort to break and level the ice for efficient and smooth travel (Druckenmiller et al., 2010). These trails, which allow hunters to haul their boats and whaling equipment to their camps using snowmobiles and sleds, also represent the hunters' escape routes to land when faced with unsafe conditions on the ice. A hunting crew's inability to quickly retreat to land or to safer ice during a break-out of the shorefast ice has often resulted in hunters drifting out to sea, without much control over their fate. Throughout history, many lives have undoubtedly been lost in this way. More recently, however, loss of life has been avoided owing to the presence of a local search and rescue helicopter that can respond rapidly to distress calls over radio (George et al., 2004b). Yet, any experience of drifting out to sea on a crumbling ice floe leaves a hunter changed forever, imprinting to memory the dangers of hunting from sea ice and how critical it is to listen to the knowledge of their Elders and the clues provided by the ice, wind and ocean.

In 2001, Warren Matumeak (1927–2010), an Utqiaġvik hunter and past director of the NSB DWM, worked with Craig George to sketch the first documented map of Utqiaġvik spring ice trails, carefully noting the various ice types that the hunters encountered (see Fig. 4.4). Scientists typically distinguish ice by thickness, roughness and age, while hunters have many more defining characteristics; for example, those addressing strength, suitability for hunting and travel, or the behaviour of animals. As an outcome of accompanying Eicken to Utqiaġvik in spring 2006 and meeting Craig George, the concept of mapping and monitoring Utqiaġvik's ice trails eventually developed into the topic of my doctoral studies in geophysics (see Druckenmiller, 2011). Over several years (2006–2011), I coordinated the mapping project with the BWCA, the NSB DWM and local whaling crews to document seasonal and inter-annual variability in the routing and condition of Utqiaġvik's trails, combining local observations with satellite imagery and reconciling hunters' assessments of ice conditions with ice thickness data from geophysical surveys (Druckenmiller et al., 2013). The project was very effective in generating interest from the community, owing in large part to my close collaboration with biologist Craig George and local hunter Lewis Brower, who assisted in communicating the role of the project to hunters and in creating maps for the whaling crews and the local search and rescue office. We conducted interviews with over a dozen hunters to discuss the suitability of sea ice for hunting, risks encountered throughout the spring season and their strategies for

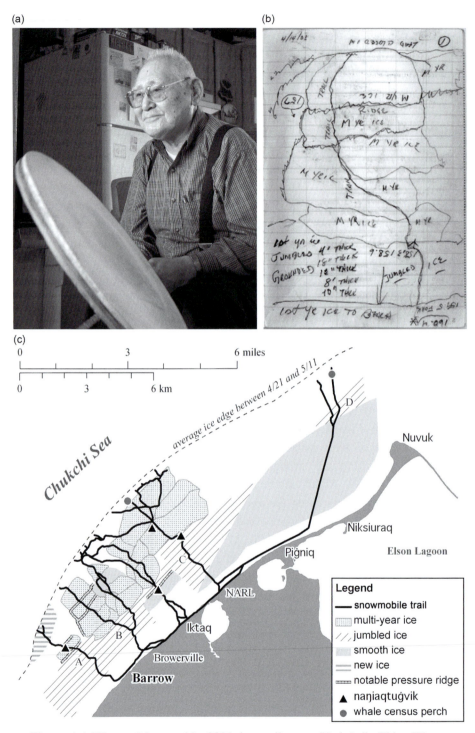

Figure 4.4 Warren Matumeak's 2001 ice trail map. Utqiaġvik Elder Warren Matumeak (top left; photo by Shari Gearheard) sketched Utqiaġvik's first map of the hunters' ice trails (top right). His maps, which were later digitized with the help of GPS tracks collected by Craig George, denoted important ice types and the locations of whaling camps (*naŋiaqtuġvik*). See Druckenmiller (2011)

safe and respectful whaling (Druckenmiller et al., 2010). Most of these interviews have since been archived by UAF's 'Sea Ice Project Jukebox' and are freely accessible by the public via the Internet.

The topic of ice safety emerged as a successful focus for knowledge sharing between local experts and scientists for several key reasons. First, the topic is vitally important to the community. During spring whaling, several hundred village members may be on the ice simultaneously, especially when the community joins together to haul a caught whale onto the ice for butchering. Large-scale ice break-out events have occurred several times in recent history (George et al., 2004b), involving many of the local hunters who have collaborated with the NSB as well as visiting scientists. Luckily, thanks to modern search and rescue resources, these recent break-out events have not claimed human lives.

Second, the topic of ice stability directly connects to a focus of the hunters' local and traditional knowledge and to personal interests, especially for those highly active hunters who venture onto sea ice throughout the year, intimately observing the evolution of the ice throughout the seasons. For those experts, the safety and stability of the ice is an ongoing assessment that revolves around first-hand encounters with very specific ice features and a timeline of the development of ice, which unfolds uniquely each year. Iñupiat sea ice knowledge is highly nuanced and, because so much of it relates to personal experience, varies between individuals. Experienced hunters are often viewed as the most cautious and talented observers, and those with a tendency for foresight (Nelson, 1969, pp. 124, 376–377). Yet, no hunter can ever predict exactly when shorefast sea ice will fail. The key to local experts' safety on ice is recognizing specific environmental conditions during which they must be on alert. They rely on indications from the wind, currents, marine life and the ice itself to decide when the risk is too high and they must return to the coast.

Third, knowledge sharing has been successful due to a diverse and interdisciplinary group of scientists and the community engaging with each other around the topic of ice hazards. To a hunter, questions pertaining to sea ice are usually about much more than ice. For example, an inquiry from a scientist into whether the ice is safe to travel on requires context and consideration of multiple variables. The safety and stability of the ice is considered relative to wind direction, strength of ocean currents and also in terms of how long or how far one may wish to travel on the ice. Safety is often viewed in the context of how much time one may have to get off the ice in response to an environmental warning sign. By engaging with hunters over many years as an informal but connected network of scientists from various disciplines (geophysics, oceanography, biology and the social sciences), we achieved an ongoing conversation about ice that was compatible with their holistic, experiential and situational knowledge and perspectives.

These relationships have allowed certain ice-related topics to remain the basis for continued and revisited discussions over many years. A single conversation is rarely enough to bridge the divide across two very different ways of knowing and where there is not always common terminology. On several occasions, while we discussed ice conditions, a hunter, such as Lewis Brower, would point out that what they were currently telling me had been told to me before on several previous occasions, albeit perhaps in a slightly different context. The holism and continuity of their worldview is replicated in how they share their knowledge, no matter how fragmented it may sometimes appear as a result of separate, seemingly disjointed conversations. Indigenous experts often possess a remarkable ability to recall past discussions and moments of sharing, which indicates the importance they attribute to the sharing process and to the knowledge being shared.

For local hunters on the North Slope, discussing ice conditions in the context of hunting whales, seals or walrus is a very natural discussion to have in a coastal Arctic community. It is a timeless topic that allows hunters to connect with the knowledge and experiences of their ancestors. With observations of a changing climate, there is recognition that conditions today are different from those of the past; however, so many pieces remain fundamentally unchanged – the people, the animals, the ocean and the need to harvest to feed the community. When the focus of knowledge sharing shifts towards recent and emerging issues that were once unfamiliar to the Arctic, such as the adverse effects of oil and gas development or the potential for oil spills in the ocean, the tone of knowledge sharing perceptively changes and the perspective pivots towards an uncertain future. Yet, it is clear that contributing their knowledge to such topics is a priority.

Workshops on Coastal Currents, Ice and Emergency Response

The application of Indigenous knowledge to assess and mitigate the environmental impacts of natural resource development is not new (Huntington, 2013). One of the earliest examples in Alaska was the incorporation of Iñupiat traditional knowledge into the Environmental Impact Statement for the Northstar oil prospect in the Beaufort Sea (US Army Corps of Engineers, 1999). Years later, the engagement and documentation of traditional knowledge in the context of oil and gas development became a focus of the Bureau of Ocean Energy Management (BOEM, 2012), the Environmental Protection Agency (e.g., Stephen R. Braund and Associates, 2011) and of individual oil production companies. In 2010, Shell Exploration and Production Company[3] and the NSB signed a cooperative research agreement to conduct baseline studies regarding the Arctic marine ecosystem and the vulnerability of subsistence resources and hunting to oil and gas exploration activities. The primary goal was to increase the knowledge base for informed development decisions (Baseline Studies Agreement, 2010). A science steering committee was established, comprising two Shell representatives, three independent scientists including Hajo Eicken as a sea ice geophysicist, two NSB DWM biologists including Craig George and local Indigenous experts from each of the six North Slope coastal villages (Utqiaġvik, Kaktovik, Nuiqsut, Point Hope, Point Lay and Wainwright; see Fig. 4.1). With funding from Shell, the science steering committee oversaw the NSB DWM's implementation of a broad collection of projects, most or all of which incorporated local and traditional knowledge.

In 2013, the NSB/Shell Baseline Studies Programme supported a collection of workshops to explore the state of knowledge on the role of coastal currents, local weather patterns and bathymetry in controlling ice movement and the dispersal of marine life in the north-eastern Chukchi Sea. The idea for these workshops, however, developed during an earlier meeting in Utqiaġvik in 2008, focused on improving stakeholder information exchange relevant to coastal and offshore oil and gas exploration (Haley and Eicken, 2011). At this meeting, it was recognized that critical knowledge gaps existed in the information necessary for effective emergency response in coastal, seasonally ice-covered waters, which are often poorly understood owing to limited in-situ and remote sensing observations.

Following the determined efforts by Hajo Eicken and others, the recommended workshops took place in March 2013 in the villages of Utqiaġvik and Wainwright, with participation from the villages of Nuiqsut and Point Lay. Each workshop lasted two days with approximately twenty-five participants each. The respective Whaling Captains Associations for each village largely determined who from the communities would attend. The identified local experts were highly experienced Elders, active hunters and community leaders with familiarity of current emergency response capabilities. The participating scientists came from the NSB DWM, UAF, the National Snow and Ice Data Center, NOAA's Office of Response and Restoration, and Shell Oil Company. The scientists represented a mix of those with long-standing collaborative ties to the communities and first-time visitors to the North Slope hoping to establish new personal connections. For several first-timers, this was 'initiation training' in Indigenous knowledge and exposure to a problem-focused knowledge co-production effort.

The ultimate goal of the workshops was to identify opportunities to improve coastal environmental monitoring, ocean and oil spill trajectory modelling, and risk management and emergency response capabilities on the North Slope. Throughout the workshops, the approach generally focused on where critical data and knowledge gaps exist and where different bodies of knowledge agree, disagree or may complement each other. The discussions focused on the dominant, persistent or recurring spatial patterns that are observed in coastal waters, either by hunters or scientific instrumentation, such as high frequency coastal radars

(Statscewich et al., 2011). Based on years of previous discussions, the scientists were interested in exploring local knowledge of where marine mammals aggregate and where floating material typically washes ashore. This information indicates important coastal ocean processes relevant to predicting where spilled oil or a damaged ship might also be found washed-up along the coastline and, therefore, could identify where to prioritize coastline protection and place skimmers to collect oil in the event of a spill.

Local experts, like scientists, generally seem to understand the boundaries of their knowledge, but it is rare that the two have the chance to explore their limitations together. Scientists often possess the physical or ecological under-standing of a process that can apply at a regional scale, such as large-scale atmospheric patterns, controls on the strength of the Beaufort Gyre (i.e., the predominantly clockwise ocean circulation within the Beaufort Sea) or the implications of increased river run-off and sea ice melt on the biological properties of Arctic waters. In contrast, local experts have a much clearer understanding of how regional processes manifest at the local scale, and accordingly discuss such topics in the context of their everyday hunting and travel experiences. Iñupiat whaling captains and other experienced hunters from Wainwright, Utqiaġvik and Nuiqsut described the waters off North Slope communities in terms of the relative strength of ocean currents and how they converge, move ice, and are indicative of and respond to changing winds. They identified locations where coastal rip currents occur and where opposing currents come together to increase the choppiness of the water's surface (Fig. 4.5).

Several hunters from Utqiaġvik discussed the critical limitations of navigating coastal waters during fall freeze-up, when launching and operating a boat is made difficult by slush ice formation, lagoon freeze-up and shoreline freezing. At this time of year, hunters may still be pursuing bowhead whales during their autumn hunt within open water or already venturing into whatever ice is nearby to hunt seals. This time of year also happens to be when offshore and near-shore industrial operations may still be withdrawing their marine-based equipment in preparation for the oncoming winter. The arriving autumn ice can present unforeseen and challenging obstacles for such ill-timed withdrawals. With their keen ability to perceive slight changes in ocean conditions, local Iñupiat experts demonstrate a body of expertise that could prove critical to any necessary emergency operations, particularly when conducted from small boats efficiently suited for rapid deployment and navigation through coastal lagoons.

Practical outcomes resulted from what amounted to four days (two days in each village) of lengthy engaged discussions, all indoors, around a table and over much coffee. The workshops recognized the need for more focused future studies to compare community-identified convergence locations and drift patterns with the

Figure 4.5 North Slope Borough biologist Craig George (centre) and Utqiaġvik whaler Margaret Opie (right) discussing the location of the main coastal ocean currents off Utqiaġvik during the 2013 workshop. [A black and white version of this figure will appear in some formats. For the colour version, please refer to the plate section.]
Photo by Nokimba Acker

results of coastal radar data and oil spill trajectory models. Participants also highlighted the need to identify shoreline staging locations for emergency response efforts and safe marine access locations for navigating within barrier islands and lagoons, especially where nautical charts are currently inaccurate owing to migrating coastlines. These are two of many recommendations included in the workshop report (see Johnson et al., 2014).

The most meaningful outcome was perhaps the recognition that attempts to compare knowledge and observations are very often limited by the lack of correspondence between Iñupiat and scientific terms and concepts. It can be challenging for local experts and scientists to know whether they are actually referring to the same ice and ocean features when they use different terms and lack the shared experience of observing specific features together in the real-world environment. The extent to which local Indigenous experts can contribute to improved information for coastal emergency preparedness and response, which is largely if not exclusively framed in scientific terms, relies, in part, on establishing this correspondence. Without it, it is difficult to move towards knowledge sharing that has practical application. For example, in the event of a major spill or incident, aircraft over-flights would likely assess the extent of the spill and identify key locations where a focused response would be needed. However, in the earliest moments following an event and when conditions are not safe or suitable for over-flights, local boats captained by community members could prove critical to a successful response, providing that a protocol and common terminology were established.

At the workshops, there was uncertainty about the equivalence between the visible ocean boundaries described by hunters and oceanographers (e.g., see Fig. 4.6d). *Yuayuk* in Iñupiaq describes 'where two currents meet and where the ocean surface is choppy', often identified by the presence of aggregating bowhead whales or *ugruk* (bearded seals). *Yuayuk* is comparable in meaning to what oceanographers refer to as a 'front', which is a boundary between opposing currents that typically separates waters of distinct physical characteristics (temperature, salinity or density), or more specifically, a shear or transition zone, which would describe, in oceanographic terms, the marking of a boundary between currents of differing velocities. When terminology is comparable in general meaning, there is often the more difficult question of whether the words would be used to describe the same ocean feature if jointly observed first hand by scientists and Indigenous knowledge holders. During the following summer, Hajo Eicken and UAF oceanographer Mark Johnson, who both participated in the workshops, joined local experts from Utqiaġvik to co-observe surface ocean features together by boat (see Fig. 4.7). This represented an important step in pursuing mutual understanding even though they were not able to fully reconcile their usage of terminology for current boundaries within a single day of boating. To this day, Eicken and Johnson continue to work with the community of Utqiaġvik to better understand near shore ice–ocean interactions.

Richard Glenn, an Iñupiat whaler who also has a graduate degree in geology from UAF, co-facilitated the workshops. He ended the meeting in Wainwright by reminding participants of not only why they were there but also why they were chosen. His words emphasized the importance of specific individuals and the relationships they uphold with the local environment and with other experts.

You were selected because you're out there; you're the eyes and ears outdoors; not because you're the head of anything or have any other constituency other than reality. And that is something that is respected, and respected by the researchers who came here to talk with you. Similarly, I respect them, because they're the ones with dirty boots. They've been out in our ocean. They've been on our beaches, on our ice. So, I think we have a good group of folks talking to each other. I hope that what we've done for the past two days is useful and continues.

Richard Glenn, Iñupiat whaler and geologist

(a)

(b)

(c)

(d)

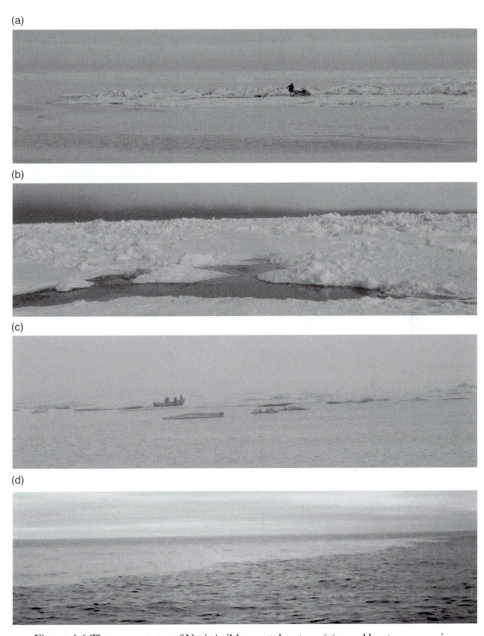

Figure 4.6 The many states of Utqiaġvik's coastal waters: (a) a seal hunter on new ice in fall; (b) a deteriorating ice trail in spring with a 'water-sky' (clouds indicating open water) on the horizon; (c) boating in summer fog among drifting ice floes; and (d) a shear boundary between opposing currents during the open water season. [A black and white version of this figure will appear in some formats. For the colour version, please refer to the plate section.]
Photos (a), (b) and (c) by Matthew L. Druckenmiller; Photo (d) by Hajo Eicken

Figure 4.7 Utqiaġvik hunter and ice expert Billy Adams (left) and UAF oceanographer Mark Johnson (right) discussing ocean currents while boating in the waters off Utqiaġvik.
Photo by Hajo Eicken

These workshops provided a lens through which to view past success in knowledge sharing during the prior decades of scientists cooperating with North Slope communities. Existing relationships between key individuals were instrumental in encouraging the involvement of a greater numbers of scientists and local experts. With this basic foundation for sharing recognized, these workshops provided a meaningful opportunity for additional reasons. First, through trust, we were able to avoid emotional responses to what is a very sensitive topic in Arctic Alaska, responses that can severely derail well-intentioned meetings. The Exxon-Valdez Oil Spill in 1989 provided vivid images for all Alaskans of the damage that such events can have on ecosystems and communities. Second, a focus was placed on the importance of terminology, both scientific and Iñupiaq. A focus on Native words both honours the language and provides a clear entry point for sharing. Lastly, relationship building was a clearly articulated goal for the workshops at the outset, rather than an additional benefit recognized later. Afterwards, several village residents indicated that the most valuable outcome of the workshops was that they established communication channels with knowledgeable scientists and emergency response professionals, who may have specific roles in responding to local emergencies.

The workshops, together with the preceding meeting in 2008 (Haley and Eicken, 2011), successfully demonstrated a 'community of practice' – the coming together of experts and practitioners from various backgrounds to contribute their knowledge and perspectives to solving problems within a well-defined domain (Wenger et al., 2002). Eicken et al. (2011) identified a community of practice as an effective means of addressing issues related to offshore oil and gas development on the North Slope where local peoples are not fluent in current offshore technologies (Glenn et al., 2011); the technical and logistical capabilities of industry to respond to coastal emergencies; or how their local and traditional knowledge could potentially merge with formal preparedness planning or response. Collaboratively addressing these knowledge gaps assists local communities in shaping their long view on the sustainability of increased activities in the Arctic and where knowledge co-production is needed in the years ahead.

Co-Producing Knowledge for Use: A Message for the Arctic Ice and beyond

Dale and Armitage (2011) define knowledge co-production as involving 'a plurality of knowledge sources and types brought together to address a defined problem and build an integrated or systems-oriented understanding of that problem'. They continue to identify the five dimensions: knowledge gathering, sharing, integration, interpretation and application. Similarly, Hegger et al. (2011) suggest that knowledge co-production 'implies that scientists, policymakers and sometimes other societal actors cooperate in the exchange, production and application of knowledge'. The latter arrive at this definition through an emphasis on cooperation between scientists and policymakers (not local experts); however, the correspondence between these two definitions is the emphasis on knowledge application.

Indigenous knowledge of the Arctic peoples has found an assortment of fates within research: it has been recorded, documented in books, deconstructed and archived in databases, integrated with science and projected on maps. It has been used to characterize the livelihoods of Northern peoples, co-manage marine and terrestrial resources, and better understand the state of the changing Arctic. Such work has changed the way research proposals are reviewed and funded, led to the creation of various institutions that are intended to both foster and buffer science-community interaction, and necessitated a closer look at research ethics and intellectual property rights. In other words, Indigenous knowledge is often captured and documented as a means to inform scientists, decision-makers and managers but is still seldom put into action in a context beyond traditional use.

Collaborative research with Indigenous peoples must increasingly move beyond documenting knowledge as collections of discrete pieces of knowledge or observations towards understanding Indigenous knowledge as a process (see Berkes, 2009). 'Knowledge as process' acknowledges both the cultural and practical contexts in which knowledge is used. In a modern North Slope community, as in most Indigenous communities across the world, the use of local and traditional knowledge is blended with the use of modern tools such as handheld global positioning systems, refined forms of transportation and new forms of information sharing, such as through Facebook, YouTube and other forms of online multimedia. To view 'knowledge as process' is to acknowledge how information is shared and to ideally discover new ways and audiences for sharing. Here, I suggest that knowledge co-production is not about taking a collection of scientific information and synthesizing it with a collection of traditional knowledge to arrive at new blended nuggets of information. This happens and can be extremely valuable, but what is missing is the dimension in which knowledge systems grow into interdependence – where the knowledge 'processes' are mutually reliant on each other.

An established, well-told and respected record of scientists and Indigenous knowledge holders working together is important for the relationships that it develops and the confidence that it instils in people that their efforts are part of a larger story of cultural persistence. On several occasions, I have heard it said by a local person on the North Slope that the increasing interests and activities in the Arctic by 'outsiders' will one day recede and that Indigenous communities will remain, continuing to depend on the environment in the ways that their communities have for centuries. This is not to say that North Slope communities will not continue to evolve with new technology and globalization, but rather that they are the only constant and long-term residents of the region. Their perspectives are linked to a distant horizon, where their foundational priorities are unchanged. Whether this will prove true or not, a strengthened commitment towards knowledge co-production – the development of applied knowledge that is dependent on both science and Indigenous knowledge – will benefit the Iñupiat as they encounter a rapidly changing and increasingly interconnected world.

Acknowledgements

The collaborations described in this chapter were the results of many key individuals from North Slope communities and the science community, as well as several organizations: the North Slope Borough Department of Wildlife Management, Alaska Eskimo Whaling Commission, Barrow Whaling Captains Association, Barrow Arctic Science Consortium, University of Alaska Fairbanks and the Seasonal Ice Zone Observing Network. For the coastal current and ice workshops, I thank the communities of Utqiaġvik, Wainwright, Nuiqsut and Point Lay. Funding support for the workshops was provided by the North Slope Borough under a funding agreement through the North Slope Borough – Shell Baseline Studies Program. Supplemental support came through the North by 2020 Forum at the International Arctic Research Center of the University of Alaska Fairbanks and in-kind support through the National Oceanic and Atmospheric Administration's Office of Response and Restoration. I also thank the editors of this volume, along with Courtney Carothers and Laura Zanotti, for their helpful comments and review of this chapter.

Notes

1 Utqiaġvik, formerly Barrow, reverted to this traditional Iñupiat name in 2016. Utqiaġvik refers to a place to gather wild roots.
2 These interviews are now available as online recordings through the University of Alaska Fairbanks' Oral History Program's *Sea-Ice Project Jukebox*.
3 In autumn 2015, during the initial writing of this chapter, Shell Exploration and Production Company indefinitely pulled their offshore oil and gas operations out of Alaska, citing exploration results that did not economically warrant development. Shell was a major leading company in the pursuit of developing Alaska's offshore oil and gas reserves, and thus their absence has dramatically shifted the regional focus and priorities of the industry. The near- and long-term economic impacts of Shell's withdrawal are still being realized on the North Slope and in Alaska in general.

References

Albert, T. F. 2000. The influence of Harry Brower, Sr., an Iñupiaq Eskimo hunter, on the bowhead whale research program conducted at the UIC-NARL facility by the North Slope borough. In Norton, D. W. (ed.) *Fifty More Years below Zero: Tributes and Meditations for the Naval Arctic Research Laboratory*. Fairbanks: University of Alaska Press, pp. 265–278.

Anungazuk, H. 2003. Whaling: Indigenous ways to the present. In McCartney, A. P. (ed.) *Indigenous Ways to the Present: Native Whaling in the Western Arctic*. Edmonton: Canadian Circumpolar Institute Press, pp. 427–432.

Baseline Studies Agreement. 2010. Collaborative research agreement by and between the North Slope borough and Shell exploration & production company, NSB Contract Number 2011-080.

Berkes, F. 2009. Indigenous ways of knowing and the study of environmental change. *Journal of the Royal Society of New Zealand*, 39(4): 151–156. https://doi.org/10.1080/03014220909510568

Boas, F. 1888. The Central Eskimo. Sixth annual report of the Bureau of Ethnology 1884-1885: 399–675.

Bockstoce, J. 1986. *Whales, Ice, and Men: The History of Whaling in the Western Arctic*. Seattle: University of Washington Press.

BOEM. 2012. Special issue on traditional knowledge. *BOEM Ocean Science*, 9(2): 1–16.

Brewster, K. 1997. Native contributions to Arctic sciences at Barrow, Alaska. *Arctic*, 50(3): 277–288. https://doi.org/10.14430/arctic1109

Brewster, K. 1998. *An Umialik's Life: Conversations with Harry Brower, Sr.* M.A. Thesis. University of Alaska Fairbanks.

Dale, A. and Armitage, D. 2011. Marine mammal co-management in Canada's Arctic: Knowledge co-production for learning and adaptive capacity. *Marine Policy*, 35: 440–449. https://doi.org/10.1016/j.marpol.2010.10.019

Druckenmiller, M. L. 2011. *Alaska Shorefast Ice: Interfacing Geophysics with Local Sea Ice Knowledge and Use.* Doctoral thesis. Fairbanks: University of Alaska Fairbanks.

Druckenmiller, M. L., Eicken, H., George, J. C. and Brower, L. 2010. Assessing the shorefast ice: Iñupiat whaling trails off Barrow, Alaska. In Krupnik, I., Aporta, C., Gearheard, S., Laidler, G. J. and Kielsen Holm, L. (eds.) *SIKU: Knowing Our Ice.* New York: Springer-Verlag, pp. 202–228.

Druckenmiller, M. L., Eicken, H., George, J. C. and Brower, L. 2013. Trails to the whale: Reflections of change and choice on an Iñupiat icescape at Barrow, Alaska. *Polar Geography*, 36(1–2): 5–29. https://doi.org/10.1080/1088937X.2012.724459.

Druckenmiller, M. L., Eicken, H., Johnson, M., Pringle, D. and Williams, C. 2009. Toward an integrated coastal sea-ice observatory: System components and a case study at Barrow, Alaska. *Cold Regions Science and Technology*, 56(1–2): 61–72. https://doi.org/10.1016/j.coldregions.2008.12.003

Eicken, H., Ritchie, L. A. and Barlau, A. 2011. The role of local and Indigenous knowledge in Arctic offshore oil and gas development, environmental hazard mitigation, and emergency response. In Lovecraft, A. L. and Eicken, H. (eds.) *North by 2020: Perspectives on Alaska's Changing Social-Ecological Systems.* Fairbanks: University of Alaska Press, pp. 577–603.

Foote, D. C. 1960. The Eskimo hunter at Point Hope, Alaska: September, 1959 to May, 1960, United States Atomic Energy Commission: 58.

Gearheard, S. F., Holm, L. K., Huntington, H., Leavitt, J. M., Mahoney, A. R., Opie, M., Oshima, T. and Sanguya, J. (eds.) 2013. *The Meaning of Ice: People and Sea Ice in Three Arctic Communities.* Hanover, NH: International Polar Institute Press.

George, J. C., Huntington, H. P., Brewster, K., Eicken, H., Norton, D. W. and Glenn, R. 2004a. Observations on shorefast ice dynamics in Arctic Alaska and the responses of the Iñupiat hunting community. *Arctic*, 57(4): 363–374. https://doi.org/10.14430/arctic514

George, J. C., Zeh, J., Suydam, R. and Clark, C. 2004b. Abundance and population trend (1978–2001) of Western Arctic bowhead whales surveyed near Barrow, Alaska. *Marine Mammal Science*, 20(4): 755–773. https://doi.org/10.1111/j.1748-7692.2004.tb01191.x

Givens, G. H., Edmondson, S. L., George, J. C., Suydam, R., Charif, R. A., Rahaman, A., Hawthorne, D., Tudor, B., DeLong, R. A. and Clark, C. W. 2013. Estimate of 2011 abundance of the Bering–Chukchi–Beaufort Seas bowhead whale population. Paper SC/65a/BRG01.

Glenn, R., Itta, E., Napageak, Jr., T. and Klick, M. 2011. Local perspectives on the future of offshore oil and gas in Northern Alaska. In Lovecraft, A. L. and Eicken, H. (eds.) *North by 2020: Perspectives on Alaska's Changing Social-Ecological Systems.* Fairbanks: University of Alaska Press, pp. 605–614.

Haley, S., and Eicken, H. 2011. Introduction: Coastal and offshore oil and gas development: Balancing interests and reducing risks through collaboration and information exchange. In Lovecraft, A. L. and Eicken, H. (eds.) *North by 2020: Perspectives on*

Alaska's Changing Social-Ecological Systems. Fairbanks: University of Alaska Press, pp. 495–501.

Hegger, D., Lamersb, M., Van Zeijl-Rozema, A. and Dieperink, C. 2011. Knowledge co-production in climate change adaptation projects: what are the levers for action? Colorado Conference on Earth System Governance, Fort Collins, CO, May 17–20.

Huntington, H. P. 2013. Traditional Knowledge and Resource Development. Gap Analysis Report #11. Resources and Sustainable Development in the Arctic.

Huntington, H. P., Brower, Jr., H. and Norton, D. W. 2001. The Barrow symposium on sea ice, 2000: Evaluation of one means of exchanging information between subsistence whalers and scientists. *Arctic*, 54(2): 201–206. https://doi.org/10.14430/arctic780

Johnson, M., Eicken, H., Druckenmiller, M. L. and Glenn, R. 2014. (eds.) Expert Workshops to Comparatively Evaluate Coastal Currents and Ice Movement in the Northeastern Chukchi Sea; Barrow and Wainwright, Alaska, March 11–15, 2013. University of Alaska Fairbanks, Fairbanks, AK.

Krupnik, I. and Jolly, D. (eds.) 2002/2010. *The Earth Is Faster Now: Indigenous Observations of Arctic Environmental Change*, Second edition. Arctic Research Consortium of the United States, Fairbanks, AK: Arctic Research Consortium of the United States.

Krupnik, I., Aporta, C., Gearheard, S., Kielsen Holm, L. and Laidler, G. (eds.) 2010. *SIKU: Knowing Our Ice: Documenting Inuit Sea Ice Knowledge and Use*. New York: Springer-Verlag.

Lindsay, R. and Schweiger, A. 2015. Arctic sea ice thickness loss determined using subsurface, aircraft and satellite observations. *The Cryosphere*, 9: 269–283, https://doi.org/10.5194/tc-9-269-2015.

Lowenstein, T. 1980. *Some Aspects of Sea Ice Subsistence Hunting in Point Hope, Alaska: A Report for the North Slope Borough's Coastal Zone Management Plan*. North Slope Borough, Barrow, Alaska.

Mahoney, A. R. 2006. *Alaska Landfast Sea Ice Dynamics*. Doctoral thesis, University of Alaska Fairbanks.

Nelson, R. 1969. *Hunters of the Northern Ice*. Chicago: University of Chicago Press.

NOAA. 2016. Federal Register, Vol. 81, No. 32, February 18, 8177.

Quakenbush, L., Citta, J., George, J. C., Heide-Jørgensen, M. P., Small, R., Brower, H., Harwood, L., Adams, B., Brower, L., Tagarook, G., Pokiak, C. and Pokiak, J. 2012. Seasonal movements of the Bering-Chukchi-Beaufort stock of bowhead whales: 2006–2011 satellite telemetry results. Report to the Scientific Committee of the International Whaling Commission, SC/64/BRG1.

Statscewich, H., Weingartner, T., Danielsen, S., Grinau, B., Egan, G. and Timm, J. 2011. A high-latitude modular autonomous power, control, and communication system for application to high-frequency surface current mapping radars. *Marine Technology Society Journal*, 45: 59–68. https://doi.org/10.1109/OCEANS-Yeosu.2012.6263620

Stefansson, V. 1919. *My Life with the Eskimo*. New York: Macmillan.

Stephen R. Braund and Associates 2011. Chukchi and Beaufort seas national pollutant discharge elimination system exploration general permits reissuance: report of traditional knowledge workshops – Point Lay, Barrow, Nuiqsut, and Kaktovik. Prepared for Tetra Tech. and the U.S. Environmental Protection Agency. March 11, 2011.

Stern, P. R. 2004. *Historical Dictionary of the Inuit, Historical Dictionary of Peoples and Cultures*. No. 2, Lanham, MD: The Scarecrow Press, Inc.

Stroeve, J. C., Serreze, M. C., Holland, M. M., Kay, J. E., Maslanik, J. and Barrett, A. P. 2012. The Arctic's rapidly shrinking sea ice cover: A research synthesis. *Climatic Change*, 110(3–4): 1005–1027 https://doi.org/10.1007/s10584–011-0101-1.

Thewissen, J. G. M., George, J., Rosa, C. and Kishida, T. 2011, Olfaction and brain size in the bowhead whale (*Balaena mysticetus*). *Marine Mammal Science*, 27(2): 282–294. https://DOI.org/10.1111/j.1748-7692.2010.00406.x.

Tillman, M. 1980. Introduction: A scientific perspective of the bowhead whale problem. *Marine Fisheries Review*, 42(9-10): 1–5.

US Army Corps of Engineers. 1999. Beaufort Sea oil and gas development/Northstar Project: final environmental impact statement. 4 volumes + appendices. Prepared by the US Army Corps of Engineers, Alaska District.

Wenger, E., McDermott, R. and Snyder, W. M. 2002. *Cultivating Communities of Practice: A Guide to Managing Knowledge*. Boston, MA: Harvard Business School Press.

Woodby, D. A. and Botkin, D. B. 1993. Stock sizes prior to commercial whaling. In Burns, J. J., Montague, J. J. and Cowles, C. J. (eds.) *The Bowhead Whale*. Society for Marine Mammology, Special Publication 2. Lawrence, KS: Allen Press, pp. 387–407.

5

Indigenous Ice Dictionaries: Sharing Knowledge for a Changing World

IGOR KRUPNIK

Why Sea Ice Matters

To many people living in the temperate mid-latitude areas, 'climate change' is about heat and warming. Not so in the Arctic, where the message of climate change is more often 'it's not cold enough' (Krupnik et al., 2010c). Indigenous people, particularly those living on the seashores, depend on long cold winters to build solid offshore ice. They use it as a platform for travel and hunting for the animals that sustain their life (Druckenmiller, Chapter 4). No wonder people observe the ice, know it intimately and have many words to describe it.

The Indigenous claim 'it's not cold enough' makes perfect sense from the principles of sea ice formation. It requires cold days – many cold days – to build a solid ice cover. The ice that is built 'properly', that is, in cold and calm weather, would stay longer; sustain the summer melt; resist and reflect the heat because of its white surface; and would protect the shoreline from storms, erosion and permafrost melt. If the ice is weak and broken, comes late or leaves early, more heat is absorbed into the ocean that may produce thinner and weaker ice in the following winter. As this process goes on, it makes the sea ice a highly sensitive indicator of Arctic warming.

Over the past 40 years of satellite observations, Arctic sea ice has declined dramatically in its seasonal extent, overall volume, age and duration (Fig. 5.1; Vaughan et al., 2013: 323–330; Larsen and Anisimov, 2014; PIOMAS, 2016; Eicken et al., Chapter 3). Thirteen historical minimums of the summer Arctic sea ice extent were recorded in the last thirteen years, including one in 2019 that featured the second lowest Arctic ice extent on record (Anonymous, 2019; Lindsey and Scott, 2019). The proportion of thick, solid ice that is five years old or older has declined from 20 per cent in the 1980s to less than 1 per cent today (Perovich et al., 2018). In the Bering Sea, winter sea ice distribution has changed from a predictable system of geographically arranged icescapes into a mixing bowl of

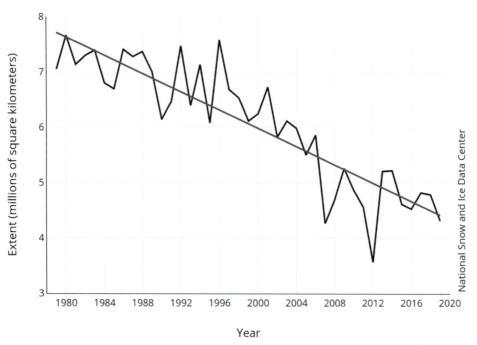

Figure 5.1 Chart of the minimal summer sea ice extent, September 1979–2019 (NSIDC – open access)

drifting floes that are easily driven to and fro by winds, storms and currents (Ray and Hufford, 1989; Ray et al., 2016).

Indigenous Arctic residents noticed transitions in their environment, including changes in sea ice, as early as the 1990s and they have spoken about it loud and clear (McDonald et al., 1997; Callaway et al., 1999; Huntington, 2000; Krupnik, 2000; Pungowiyi, 2000; Jolly et al., 2002). However, their sea ice monitoring is organized from the particular vision of users, not of ice scientists. It was, therefore, imperative to build bridges between their observations and the scientists' instrumental tracking of change: the two different systems of knowledge that use benchmarks, tools and indicators of their own (Huntington et al., 2001; Krupnik, 2002; Norton, 2002).

Another strategy is to document how Native people classify the ice and what terms they use in their local languages for its many variations and processes. This chapter introduces the study of Indigenous sea ice nomenclatures as a path towards documenting and sustaining local ecological knowledge and to 'co-produce' new knowledge about ice and Arctic change.

From 'Folk Taxonomies' to 'Sea Ice Lexicons'

The fact that the Inuit people had intimate knowledge of sea ice and many words for types of ice and snow in their languages has been known since the era of early Arctic explorations (Aporta, 2009); it was substantiated in the first published dictionaries of Inuit languages in the late nineteenth century (Erdmann, 1864–1866; Kleinschmidt, 1871; Petitot, 1876; see Krupnik and Müller-Wille, 2010). Yet, little was done to document specific Inuit terminologies for sea ice until the 1960s, at the dawn of the era of scientists' increased interest in what were then termed 'folk taxonomies' and later named 'ethnoscience' ('ethnobotany', 'ethnozoology', 'ethnomedicine': Sturtevant, 1964). All over the world, anthropologists, linguists and biologists recorded Indigenous terms for various species of birds and fish as well as edible and medicinal plants in their quest to understand how small-scale societies of hunter-gatherers and early agriculturists classified their environment (Conklin, 1955; Bulmer, 1957; Berlin et al., 1966). In the polar regions, students of folk taxonomies recorded Eskimo/Inuit names for birds (Irving, 1953, 1958; Macpherson, 1958; Höhn, 1969), plants (Bank, 1953; Oswalt, 1957; Young and Hall, 1969), and human body parts and sicknesses (Lantis, 1959). Therefore, it is all the more important to recognize the pioneering work of anthropologist Richard Nelson who produced the first list of more than eighty Iñupiaq terms for sea ice used by the North Alaskan community of Wainwright (Nelson, 1969: 398–403).

Nelson's effort to arrange the terms that he learned from local hunters and organize them according to ice age and thickness, specific forms, movement and related phenomena was an early example of 'knowledge co-production', even though Nelson did not identify his local partners and sources of his knowledge. In 2009, Ronald H. Brower, Sr., an Iñupiaq linguist originally from Utqiaġvik (formerly Barrow), re-transliterated Nelson's list in contemporary Iñupiaq orthography, with the assistance of Elders Rossman Peetook and Leo Panick from Wainwright and Jonathan Aiken from Utqiaġvik . This later effort in 'knowledge co-production' made Nelson's ice terms of the 1960s accessible to today's Wainwright Iñupiaq speakers, school students, language specialists and others.

Gambell, Alaska, 2000: Working with the Experts

The second Indigenous sea ice lexicon was compiled in the 1970s by Conrad Oozeva (1925–2016), a Yupik hunter from the community of Gambell (Sivuqaq) on St. Lawrence Island.

Oozeva recalled in a taped interview in the year 2000,

This idea [to compile a list of Yupik words for sea ice] first came to my mind while I was aboard the icebreaker 'Polar Sea'. It took place many years ago, sometime in the 1970s.

The reason I was there was because at that time there were several ice-scientists who were looking for the bowhead whales. . . . They also realized that we, Eskimos, have good eyes for ice spotting. So, they invited me to come on one of their cruises. . . . I spent a little over a month on that boat; there was nothing much for me to do. So, I began writing down the types of sea ice I knew in my own native dialect, the Yupik terms for ice. When I came back to Gambell, I already had some notes. At that time, I was also working as an instructor for the bilingual language program at the Gambell school. So, I wrote down all these types of sea ice, with their explanations in our Yupik language for the school. And I discovered that in our language there were many more types of ice than I knew of, by talking to other experienced people. . . . elderly hunters of my time

Oozeva et al., 2004: 26.

Oozeva's sea ice lexicon, '*Sikum Un'gum Aatqusluga*. Types of Marine Ice' was first reproduced in 1986 as a section in a mimeographed teacher's binder, 'St. Lawrence Island Yupik Curriculum Resource Manual' (Walunga, 1986: 23–32). It listed ninety-seven Yupik terms for various types of ice and ice conditions arranged alphabetically, with extensive explanations of up to twenty to thirty words for each entry, all in Yupik. Yet hardly anyone knew of its existence besides a small community of Yupik students, teachers and cultural workers on Alaska's St. Lawrence Island. In fact, even though the Eskimo words for different types of ice were occasionally used in the literature during the 1980s (Freeman, 1984; Riewe, 1991; Nakashima, 1993), no organized list in the form of a 'dictionary' was published in the seventeen years between Nelson's work (1969) and Walunga's curriculum of 1986.

Nevertheless, to become a knowledge source beyond a small group of Yupik users in two island communities, Oozeva's 'ice lexicon' had to undergo a 'co-production' transformation. In the year 2000, Oozeva joined our pilot project on St. Lawrence Island aimed at documenting local people's observations and knowledge of sea ice and weather change (Krupnik and Huntington, 2004: 16–21). The project team of fifteen members included several local experts of various ages. Younger, bilingual speakers included Christopher Koonooka, Christina Alowa, George Noongwook and Milton Noongwook who worked on the translation of Oozeva's Yupik list, assisted by another senior expert, Chester Noongwook (1933–2020) and Oozeva himself (Fig. 5.2). They transformed Oozeva's list of ice terms into a bilingual Yupik-English text and added remarks on distinctive meanings of certain terms in two island communities, Gambell and Savoonga. Local Yupik artist, Vadim Yenan, under Noongwook's guidance, produced some 100 pencil drawings illustrating each type of ice in Oozeva's list.

The result was a bilingual, illustrated Yupik-English sea ice 'lexicon' (Fig. 5.3; Oozeva et al., 2004: 26–53). All terms, numbering more than 100, were soon incorporated into the authoritative two-volume Yupik–English dictionary produced by the Alaska Native Language Center (Jacobson, 2008).

Figure 5.2 The author working with Conrad Oozeva on documenting Yupik sea ice knowledge (Gambell, February 2004).
Photo by G. Carleton Ray

37.   **Nevesqaghneq**

Sikut nefqaghtekat sikumun kumlasimangilnguut iwernga paninang uulsugnalghiit.

An extension of ice formed when two floes collide and stick together (G), but not get frozen to each other (S). The new extension remains steady ice; could be dangerous to work on (CO).

38.   **Ngaayuun**

Angyami atuq taana, sikum qungangani maaqellghem wata siku tepuneghminigu aastughunneghminikut ngaayuutmeng aatqeftuq tamaana. Aastughullghiini. Tawaten ngaayuutmeng atelek taana, siku qerrutaqelghii elngaatall ayaghqukaaluni uglavaghaghsigalnguq ilanganikun repall piyaqliighnalghii.

Dangerous spot (CO). This term is used when you are out hunting in the floating dense ice in a boat. When (you are) in the midst of an ice flow, and the current is massing other ice together, it is called as such. The ice is so jammed that you have to push to get the boat through, although the ice floe does no spread out easily (G). The term used for boat hunting in dense ice. Ice moving with the current from southwest to northeast passes close to other ice that is moving in the opposite direction. When this happens, the ice floes can block your way out (S).

39.   **Ngevzin**

Iighwilnguum esnengi sikullghi taakwegkut upsagitepigtaqelghiit un'gavek meghmun. Sikullghet taakwegkut ngevzitmeng aatqut. Igleghsugnaghllaget elngaatall, kiyangllu esavghalghiini neghighluku tamaagun, sikunga anigulgunghani. Uuk nalighmeng qivallugtunghani ngevzitkun tamaagun igleghtaqelghiit, naveklu piyaalghiini pinilghiit qayughllak. Sikungi taakwum sikum esneghrugi.

Smooth shorefast ice or a large ice flow whose edge is very smooth all the way towards the ocean. It is very easy to go through, and when pulling a boat on ice long ago they usually looked for that type of ice, particularly when it had snow on it. Or when there are pressure ridges on an ice floe, hunters would go to this ice because it is so good to be on when hunting by foot (G). This is similar to *kangitek*. It can be shorefast ice or a large cake of ice that is smooth and a good path for traveling (S).

40.   **Nunaavalleq**

Sikut ayveghem uugfiksaqangi, aniita sangita mingughhutesluki tagneghqwaasluki. Tamaakut nunaavallghuftut, ayveghem uugfiksaqangi.

Any form of ice floe that walrus have stayed on for a long period of time. If the walrus have stayed on the ice for five days or more, the ice will look dark (dirty) from their bodily waste (G, S).

Figure 5.3 Page from Oozeva's Yupik sea ice dictionary (from Oozeva et al., 2004: 38I)

Figure 5.4 Conrad Oozeva (centre), with Willis (left) and Nancy Walunga (right) in Gambell, 2010.
Photo by Igor Krupnik

When Oozeva worked on his list of Yupik sea ice terms in the 1970s, he was one of many middle-aged hunters in his community that also included several dozen senior men, in his words, 'elderly hunters of my time', with excellent knowledge of sea ice. Oozeva checked his terms and their Yupik explanations with these knowledgeable experts. Even in the year 2000, he and Noongwook, our other major expert, could consult with a few elderly men of their age experienced in local ice conditions and their native Yupik language (Fig. 5.4). Our team effort in the year 2000 was a form of 'added co-production' of knowledge that was then fully viable within the 1500-strong Yupik community on the island. The situation has changed in the following years because of the recent passing of many Elders including Oozeva and Noongwook, although many knowledgeable senior experts are still alive.

SIKU: Knowing Our Ice 2005–2010

The success of the St. Lawrence Island Yupik 'ice dictionary' project provided a strong incentive to expand the collection of Indigenous ice terminologies to better understand Indigenous knowledge about climate change. Simultaneously or shortly thereafter, similar efforts were launched in several Inuit communities in the Canadian Arctic (Aporta, 2003; Gearheard et al., 2006; Laidler, 2006, 2007), and Indigenous terms for sea ice entered scholarly papers and science reports (Jolly et al., 2002; Norton, 2002; Fox Gearheard, 2003; George et al., 2004; Nichols et al., 2004; Nickels et al., 2005). These terms were also featured in several museum exhibits of the time, such as 'Arctic: A Friend Acting Strangely' (at the Smithsonian Institution's National Museum of Natural History in 2006), 'Silavut: Inuit Voices in a Changing World' (University of Colorado Boulder in 2007), 'Thin Ice: Inuit Traditions within a Changing Environment' (Hood Museum of Art in 2007) and others.

To coordinate these and similar efforts, we launched an international research project named SIKU ('Sea Ice Knowledge and Use'; also *siku*, the most general Inuit word for sea ice), as part of the International Polar Year 2007–2008. It eventually developed into a collaborative programme that engaged scientists from six nations and local knowledge experts from twenty Indigenous communities in Alaska, Canada, Greenland and Chukotka in northeast Russia (Krupnik et al., 2010a, 2010b; Aporta, 2011). From 2006–2011, the SIKU team and its partners collected Indigenous vocabularies of local terms for sea ice and related phenomena from each of these regions (see the overall list and map in Krupnik and Weyapuk, 2010; Krupnik, 2011; and details on individual projects in Laidler and Elee, 2008; Laidler and Ikummaq, 2008; Johns, 2010; Laidler et al., 2010; Tersis and Taverniers, 2010; Wisniewski, 2010; Fienup-Riordan and Rearden, 2012; Bogoslovskaya and Krupnik, 2013; Fox Gearheard et al., 2013). A few examples that follow are based on my personal experience from such efforts in knowledge co-production that involved Indigenous knowledge holders.

Wales, Alaska: Before It Is too Late

One of the most notable experiences of the SIKU years was our partnership with Winton Weyapuk, Jr. (1950–2016), an Inupiat hunter and cultural activist from the community of Wales/Kingigin in Alaska (see Eicken et al., Chapter 3). In 2007, Weyapuk, then fifty-seven years of age, agreed to compile a list of sea ice terms used in his home community, provide their explanations in English and also document Kingikmiut Elders' knowledge about ice conditions in their home area. Two Wales Elders, Peter Sokienna Sereadlook (1930–2017) and Faye Ongto-wasruk (1928–2015) served as consultants, together with Herbert Anungazuk (1945–2010), who was originally from Wales, ice specialists Hajo Eicken and Matthew Druckenmiller, and linguist Lawrence Kaplan from the University of Alaska Fairbanks.

In 2007, Weyapuk drafted the first alphabetical list of 60 Kingikmiut terms that eventually expanded to more than 120 words and expressions in consultation with Anungazuk, Ongtowasruk and Sereadlook. He wrote explanations for major ice terms in Inupiaq, so that the Kingikmiut understanding of the many types of ice

Figure 5.5 Page from the 'Wales Sea Ice Dictionary' (Weyapuk and Krupnik, 2012: 54). Caption under photo: Female walrus and a calf (*isavgalik*) are resting on the ice (*nunavait*) in the midst of scattered pack ice (*tamalaaniqtuaq*) interspersed with patches of calm flat water (*quuniq*). The mass of floating pack ice, *sigu*, consists of various types of ice, such as *puktaat* – large floes, *puikaanit* – vertical blocks of ice, *qaŋiłaq* – ice floes with overhanging shelves, *taaglut* – large pieces of darker ice, and *saŋałait* – small floating pieces of dirt ice. May 21, 2007. [A black and white version of this figure will appear in some formats. For the colour version, please refer to the plate section.]
Photo and inscribed ice terms by Winton Weyapuk, Jr

would be preserved in their native dialect. He also took more than 150 photographs of various types of ice, inscribed Kingikmiut terms onto his photographs (Fig. 5.5) and provided comments on two dozen historical photographs of ice and ice-hunting in Wales taken in 1922 by biologist Alfred Bailey (Bailey, 1943), comparing historical and present-day ice conditions (see Eicken et al., Chapter 3).

The main outcome, a 112-page 'Wales Inupiaq Sea Ice Dictionary/Kiŋikmi Sigum Qanuq Ilitaavut' (Weyapuk and Krupnik, 2012), included writings by several contributors, but it is primarily a tribute to Weyapuk's dedication. Wales is a small community of 160; by 2007, it had but a handful of elderly people still fluent in their Inupiaq language. The number continues to shrink and Weyapuk's effort was more in the realm of 'salvage documentation', despite the hope expressed in his preface to the dictionary,

It is our hope that our Inupiaq words for sea ice and the English translations we collected here can help young hunters supplement what they have learned in English about sea ice in our area. Using the English translations, they may begin to understand the changing conditions as they are affected by winds and currents. It is also our hope that they can learn and begin to use some of the Inupiaq words as a way to teach those younger than themselves.

By no means is this Inupiaq sea ice 'dictionary' an attempt to revive the Inupiaq language in Wales. Only constant usage by adults and youth can achieve that. Perhaps it can be achieved through other avenues. One can only hope ...

Perhaps, this Kingikmi sea ice 'dictionary' can be viewed as a link between the way our Elders communicated in the past and today's way of communicating. ... Language, any language, is beautiful in its own way. Inupiaq, because of its construction and its concise description of the natural environment, is no less beautiful. This dictionary may help preserve parts of it ...

Weyapuk and Krupnik, 2012: 9–10.

The Wales 'ice dictionary' was the ultimate symbol of co-production as it compiled a portion of ecological knowledge preserved in an endangered language and re-introduced it in the format of a bilingual lexicon that did not exist in traditional Inupiat epistemology. It expanded Indigenous knowledge by adding contemporary and historical photos of local icescapes and hunters, addressed the future of Indigenous heritage, and promoted the role of language and knowledge documentation. Numerous copies of the dictionary were distributed in Wales among local families, school students and relatives in other communities. A beautiful book in colour, it serves as an educational tool, a symbol of local pride and now a memorial to Weyapuk himself, following his passing in 2016 (see Eicken et al., Chapter 3).

In 2013, 'Wales Sea Ice Dictionary' received an honourable mention in the 'reference' category from the Atmospheric Science Librarians International (ASLI), an international body of librarians working in the world's major geophysical libraries. This was the first time in the group's eight-year history that it honoured a book about Indigenous knowledge of climate processes, with a local expert as its first co-author: see http://blog.ametsoc.org/uncategorized/asli-chooses-the-best-books-of-the-year/

Sireniki, Russia: The Last Man Standing

As the SIKU project unfolded in 2007, I contacted Russian Yupik educator Natalya Radunovich, who lives in the city of Anadyr in northeastern Russia, and asked whether she would collaborate with her father, Aron Nutawyi, to compile a list of Siberian Yupik sea ice terms in her home community of Sireniki (Sighineq).

It was viewed as an attempt similar to Oozeva's Yupik dictionary, and I supplied Radunovich and Nutawyi with a copy of Oozeva's list as a working sample. In a few months, Nutawyi, then sixty-six years of age, compiled a lexicon of forty-eight ice terms with detailed explanations in Yupik and in Russian that Radunovich carefully transliterated. We asked Yupik artist Vadin Yenan, who illustrated Oozeva's dictionary in the year 2000, to make similar pencil drawings for Nutawyi's list that was published in 2013 (Fig. 5.6; Nutawyi and Radunovich, 2013: 72–82). Altogether, five people (six, including Oozeva's initial impact) worked to convert one Elder's memory into new 'co-produced knowledge' so that it could be shared and offered to students at local schools, as well as among linguists and ice specialists.

The reality, however, was grim. Nutawyi's version of the Yupik language is almost identical to that spoken on St. Lawrence Island. Yet in Chukotka, only a few dozen Elders still speak the Siberian Yupik, mostly women (Morgounova Schwalbe, 2015) whose expertise commonly extends to domains other than sea ice. In Nutawyi's home community of Sireniki, he was the only senior man fully fluent in Yupik, followed by a handful of middle-aged hunters with only a limited knowledge of the language. Unlike Oozeva and Weyapuk, Nutawyi could rely neither on knowledgeable peers nor on elderly experts. No wonder, his personal list of ice terms was less than half the length of Oozeva's, even if for the same language and similar ice environment.

The message is clear: a certain critical number of Indigenous speakers and knowledge holders is required to support a successful effort in knowledge co-production. Although today's younger generation of St. Lawrence Island Yupik speaks primarily in English, there is still a robust cohort of several hundred elderly speakers who maintain the traditional cultural lexicon, including many terms for sea ice. For the Siberian Yupik, such words are only retained in Elders' memories and they are quickly fading. Nutawyi passed away in 2010 and younger hunters in his native town operate with a handful of Russian or Russian-based ice terms (Yashchenko, 2016) to describe local icescapes, for which their peers on St. Lawrence Island have many dozen specific Yupik words.

Naukan, Russia: The 'Lost' Icescape (Memory-scape)

In certain cases, sea ice terminologies could be collected among people who have moved from their home areas and thus recount their former icescapes by memory, such as the former residents of the community of Naukan (Nuvuqaq) on the Russian side of Bering Strait, near the cliffs of Cape Dezhnev (East Cape), facing Alaska and the Diomede Islands across the strait (Fig. 5.7). In 1958, under the centralized policy of governmental relocations, Russian authorities closed the

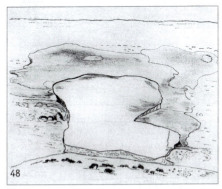

47. Мыӽа́ӄ (meghhaq)

Небольшое разводье, свободное ото льда. Хорошее место для охоты на нерпу, лахтака.

Мыӽа́ӄ сику́ӈилӈуӄ. Кия́ӈ сику́вагыт аку́ляӈитнъи. Ныӽса́ӽтулѓи.

48. Ы́лъту́гӣ́ныӄ (eltughneq)

Льдина, застрявшая у берега на подводных камнях. К ее кромке намерзает новый лед. Опасен для ходьбы из-за неизвестности и внезапных изломов.

Сику ылъту́ӄаӄ снам сны́ӈани алӈу́нак ыӽӄи́таѓискаӄ ӄалмы́сятхун. Уви́таӈа ылъту́гӣ́ны́ъм сику́ӈаӄуӄ киӈу́ваӄун. Иглы́ӽфигу́люни яку́гӣ́наӽтуӄ ӄаю́ӽлъяк каги́мьюгу́маӄ ся́ма налъю́наӽту́ӄ.

Пресный лед

49. Иӈля́ӄуӄ (сирен.), иля́ӈӄуӄ (чапл.) (inglaquq, ilangquq)

Пресный лед, кусок льда.

Мыпи́к кумля́скаӄ, иӈля́ӄупик (уӈази́г'мистун «иля́ӈӄуӄ»).

50. Иӈля́ӄуѓрак (сикўа́г'аӄ) (inglaqughraak/ sikwaghaaq)

Тонкий, только замерзающий пресный лед.

Угмы́ста́ӽаӄ нута́н кумля́лъӽи мыпи́к.

51. Иӈля́ӄулъыӄ (inglaqulleq)

Обледенение, намерзание пресного льда.

Иӈля́ӄуӈы́лг'и кумля́лъӽи мыпи́к.

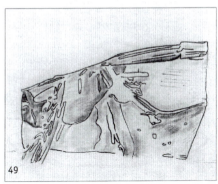

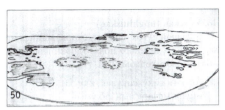

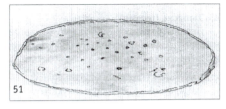

Figure 5.6 Page from Aron Nutawyi's sea ice dictionary (from Bogoslovskaya and Krupnik, 2013: 82)

Figure 5.7 Map of the Bering Strait communities (prepared by Matthew L. Druckenmiller)

village of 350 residents, with their distinct language, the Naukanski Yupik, and removed them to nearby communities (Krupnik and Chlenov, 2007). Most of the former Naukan people never had the chance to see their home area again. The younger generation speaks primarily in Russian, and children are raised in mixed Russian-Yupik and Yupik-Chukchi families.

In 2007–2008, we invited two Naukan cultural experts, Boris Alpergen (born 1941) and Elizaveta Dobrieva (born 1943), to compile a Naukan 'sea ice dictionary' for the SIKU project. They were both born and raised in Naukan and now live 100 km (60 miles) away, in the district hub of Lavrentiya. They soon prepared a list of some ninety terms and expressions, in consultation with a few elderly kinsmen. Alpergen and several younger residents of Lavrentiya produced pencil drawings to illustrate the list published in 2013 (Dobrieva and Alpergen, 2013: 148–153). Altogether, nine people collaborated on this project, including Valentina Leonova, a Yupik educator from a younger cohort, linguists Michael Krauss and Stephen Jacobson at the University of Alaska Fairbanks, the late Russian biologist Lyudmila Bogoslovskaya and myself.

We can only admire the Naukan Yupik Elders for their commitment to the cultural heritage of their former community (see Leonova, 2014). Yet after fifty years, human memory loses many details of the home landscape and icescape. Several terms in Dobrieva-Alpergen's list were translations from the much larger St. Lawrence Island lexicon. In other cases, the old Naukanski Yupik word was most obviously lost and the compilers inserted the same term for various types of ice that have specific names in the St. Lawrence Island Yupik nomenclature. Despite its decent size, the Naukanski Yupik sea ice dictionary is merely an alphabetical lexicon and it could not provide practical tips on orientation, behaviour and safety on ice.

Even when people remember the old words for sea ice and certain elements of their former habitat, they have no ability to update their knowledge of a 'historical' icescape (memory-scape[1]). They cannot detect new types of sea ice in their home area, the disappearance of the old ones and the changing pattern of ice formation under the impact of climate change. To the Elders, the icescape of Naukan remains the same, as they remembered it when they were forced to leave in 1958 (Fig. 5.8). It raises a critical issue of the role of 'living' memory in the co-production of knowledge. In an icescape that is actively used, people not only lose or replace the old terms but also add new ones that reflect the changing nature of the ice, new safety risks, new technologies, navigation and orientation skills (snowmobiles, GPS, etc.). On the contrary, a 'memory-scape' remains frozen in time and, unless

Figure 5.8 Drawing of Naukan seascape (by Boris Alpergen: from Bogoslovskaya and Krupnik, 2013: 154)

people return to Naukan or start visiting it on a regular basis, co-production would be replicating the 'old' knowledge, even in a diminishing way, with the passing of remaining elderly experts.

What Can We Learn from Indigenous 'Ice Dictionaries'?

Most of the Indigenous sea ice terminologies collected between 2005 and 2010 have now been published (Laidler et al., 2007; Krupnik et al., 2010a; Krupnik, 2011; Fienup-Riordan and Rearden, 2012; Bogoslovskaya and Krupnik, 2013; Fox Gearheard et al., 2013) but their analysis is still ongoing. We may point to certain critical lessons learned for the general process of 'knowledge co-production' in the studies of Arctic climate change.

Indigenous versus Scientific Environmental Terminologies

It is a common saying that the Inuit (Eskimo) have '200 words for snow'. It may be an old joke and an exaggeration (see Krupnik and Müller-Wille, 2010) but in many Arctic communities 40–60 Indigenous ice-related terms are often in active use and elderly experts can name up to 100–120 terms and expressions in their native language or dialect (Aporta, 2003, 2009; Laidler, 2007; Fox Gearheard et al., 2013). A critical element in the co-production of knowledge is that Indigenous ice terms often carry more information than their analogues in scientists' nomenclatures. The latter are usually framed for observers at coastal stations, on ships' bridges or flying aircraft, both in the Arctic and in Antarctica. This specifically relates to the notion of ice safety and the history of ice formation.

For example, the scientific sea ice nomenclature defines 'rotten ice', one of the forms of spring ice, as 'the melting sea ice which has become honeycombed and which is in an advanced stage of disintegration' (Arctic and Antarctic Research Institute, n.d.). A Native explanation of a similar type of ice known as *aghulleq* in Siberian Yupik is 'the old ice thinned by spring warming; extremely dangerous for walking, pulling boats or any work, even dog-driving. While walking on this ice, one always has to use a special ice-stick (*tuvek*) with a sharp iron or bone edge and continuously check the ice thickness and sturdiness for safety.' (Nutawyi and Radunovich, 2013: 73). The added value of the Indigenous definition is that it comprises practical information for local users, including on safety, traveling and forecasting.

Resilience of Cultural Icescapes

Scores of Indigenous ice terminologies that were collected between 2000 and 2010 constitute a critical dataset to assess the status of local icescapes and associated cultural knowledge at the beginning of the twenty-first century. This collection also underscored both the resilience and vulnerability of aboriginal icescapes in an era of rapid social and environmental change. To this day, many experienced hunters and Elders possess practical skills of 'reading' and forecasting the sea ice. They are competent in traditional ice nomenclatures and abide by the age-old safety rules when hunting and moving over the ice.

Even in the communities where younger and middle-aged people have already switched to English or Russian in their daily communication, many traditional practices of icescape use are sustained, such as ice fishing and seal hunting. In coastal towns and hamlets across the Arctic, where people's livelihoods are based on marine subsistence activities, people maintain a version of the local 'cultural icescape', even if in a modified or rudimentary form. New ice-related activities are often added, such as dog races, soccer games and school 'heritage' classes conducted on ice (Taverniers, 2010; Bogoslovskaya and Krupnik, 2013). They help sustain people's practical knowledge and attachment to the sea ice and thus preserve a cultural 'icescape' filled with terms, stories and names of places (Heyes, 2011).

A crucial point for knowledge co-production is that people across the Arctic still spend a lot of time on sea ice and will continue to do so in the future. Their knowledge is being preserved and transmitted within their home communities and is sustained independently of academic research or government funding for climate monitoring.

Above all, Arctic ecosystems are remarkably diverse. They often have individual microclimate and ice regimes and comprise highly distinctive sets of physical and topographic features. Local people meticulously track many signals of change in their land-, sea- and icescapes. They carefully assess the shifts in familiar weather and ice patterns; increase in storm frequency; changes to local biota and seasonal cycles of marine and terrestrial species; and a myriad other indicators, as revealed in many recent studies (McDonald et al., 1997; Riedlinger and Berkes, 2001; Krupnik and Jolly, 2002; Huntington and Fox, 2005; Nickels et al., 2005; Hovelsrud et al., 2011; Salomon et al., 2011; Fienup-Riordan and Rearden, 2012; Johnson et al., 2016). Indigenous peoples' monitoring of their home habitats is a continuous occupation that relies on hundreds of observational sites around communities, family cabins, fishing camps and along hunters' seasonal tracks. It is too precious a pool of knowledge to be missed or ignored.

Co-Production and Knowledge Loss

People's knowledge about sea ice is not immune to general trends in culture development. As Arctic residents switch from Indigenous languages to English,

Danish, Swedish or Russian, Indigenous nomenclatures for ice, snow, weather and winds fall out of use, so that younger generations often operate with a system of 'creolized' names for basic types of ice compared to many dozen traditional words that their grandparents used in the same icescape. As elderly hunters continue to preserve traditional terms that are not familiar to the youth, generational cultural gaps, 'cracks in the knowledge' (see Heyes 2011), expand. Often, the days of old Indigenous terminologies are literally numbered, as in the case of Nutawyi's list in his home community of Sireniki in Russia.

The situation is even worse for many historical icescapes (memory-scapes) of the places that were abandoned in the 1950s and 1960s. Barely a few Elders still recall the ice and associated terms in their former home areas that they were forced to leave decades ago.

Several ice nomenclatures collected for the SIKU database are 'endangered', particularly in Chukotka, Russia, Northwest Alaska, Labrador-Newfoundland and the Inuvialuktun area in Canada (Krupnik, 2011). They may not be accessible to the next generation of speakers and researchers. The lesson for the co-production is that without a consistent effort to strengthen community-based transmission, we may see a rapid transformation of local cultural knowledge by the middle of this century. By that time, the ice may still be used, but it will be described in terms from other languages common to scientists, airmen or navigators from the passing ships. There may not be much of a field for co-'production' to speak of.

In today's world of government schooling, multiculturalism and social media, three forces work to protect Indigenous knowledge: the number of speakers (or users); good channels of communication within communities; and the practical value that such knowledge reflects. As the SIKU stories illustrate, the numbers still act as the most effective safeguard, at least in the case of St. Lawrence Island Yupik. Thanks to many dedicated efforts, communication in endangered languages and knowledge systems has improved, particularly for younger speakers, so that in the words of Winton Weyapuk, we may hope that one day 'they can learn and begin to use some of the Inupiaq words as a way to teach those younger than themselves'.

Nonetheless, while the Elders work hard to preserve the words for 100 types of Arctic ice, they cannot protect the ice itself that these words describe. Their intimate knowledge was developed to navigate the ice that is not there anymore or has been changing rapidly, before their very eyes. Hunters in many communities report that they have not seen the thick bluish multi-year ice of their youth in several years and that they are forgetting the word for it. Younger people may not remember it at all (Druckenmiller et al., 2019). This word is still known in the North Slope Iñupiaq dialect (*piqaluyak*: MacLean 2014: 1184), but its days may be also numbered.

Conclusions: A Co-production for Whom?

It is common to argue that co-production produces new 'shared knowledge' by bringing together in one way or the other two (or more) different systems of knowledge (Stegmaier, 2009; Pohl et al., 2010; Hegger et al., 2012; see: Roué and Nakashima, Chapter 1). At the most general level, the co-production generates certain new paradigms and new formats; but very often they display visible birthmarks of their parental knowledge systems. Many tools used in the co-production process and, specifically, in the preparation of Indigenous sea ice lexicons, such as the concept of 'cultural icescape' or an alphabetical bilingual 'dictionary', had no roots in Indigenous epistemologies. They come directly from the 'science' world. Nonetheless, even if a bilingual lexicon of ice terms illustrated by pencil drawings, historical photographs and with added comments by ice scientists has no prototypes in Indigenous culture, it reflects the understanding of ice built by generations of local users and in their appropriate cultural terms.

Whereas the boundaries of 'co-produced' knowledge may be limited, the co-production vastly 'expands' the prospective audience for various types of knowledge and data. Even when coming from different 'paternal' epistemological systems, the products of co-production appeal to many parties and engage many constituencies in a collaborative process. A lexicon of traditional sea ice terms, even if it reflects a shrinking pool of users – as in Wales or of a single person, such as Nutwayi in Sireniki – provides an insightful resource to younger hunters, Indigenous language activists, teachers, students and scholars from many disciplines.

Co-production creates a dynamic space that generates products across established boundaries. Compiling an 'ice dictionary' is always a collaborative process that requires several partners. It engages local knowledge experts, linguists, translators and recorders. It cannot be done without the input of ice scholars who convert the Indigenous vision of ice features into terms in another language/s that may then be matched with scientific nomenclatures. It adds local artists and photographers, as well as museum and archival workers, who supply historical photos of icescapes, and editors and book designers. No wonder that making an Indigenous 'ice dictionary' commonly engages diverse teams.

A typology that I first learned from Gunn-Britt Retter, a Norwegian Sámi activist and the Head of the Arctic and Environmental Unit of the Sámi Council, offers a useful tool to assess the results of co-production. In her talk at the international conference 'Knowledge to Action' (Montreal, April 2012), she argued that it is imperative to distinguish 'for whom' a certain type of knowledge is being generated. She identified 'knowledge for science' that is created and primarily used by scientists and the environmentalist community; 'knowledge for the use of resources', whose prime producers and recipients are companies and government agencies; and 'knowledge for home'. She defined the latter as the knowledge created by Arctic Indigenous people for 'sustainable and safe life' in their home environment. By its origin and purpose, Indigenous sea ice terminologies, like other parts of LTK, belong to the category of 'knowledge for home'.

It does not necessarily mean that co-production involving Indigenous knowledge has to be practised for the sake of Indigenous partners or by Indigenous people only ('knowledge for home'). Many stories presented in this book illustrate that co-production can be equally successful when driven by scientists and/or when it produces significant new 'knowledge for science'. It is critical, though, that in many (most?) of its formats, co-production strengthens a new ethics of 'respectful coexistence' of different types of knowledge, rather than what is widely called 'knowledge integration', which involves a rather distinct approach and philosophy (Krupnik and Bogoslovskaya, 2017). Such a humanistic vision of 'respectful coexistence' should guide our co-production work involving Indigenous knowledge holders, and, at the more general level, any cross-disciplinary effort that requires partnership among physical, natural and social scientists, as well as humanists and Indigenous experts.

Acknowledgements

I am grateful to many partners in the SIKU project named in this chapter, as well as to other people with whom I have worked in the field of Indigenous knowledge documentation and co-production over many years. Marie Roué and Douglas Nakashima kindly invited me to the workshop on knowledge co-production in the Arctic (Paris, November 2014) and inspired many ideas presented in this chapter. They, as well as Shari Fox, Matthew Druckenmiller and Henry Huntington offered encouragement and valuable comments on an earlier draft. A version of this chapter was first presented as the third 'Will Morrison Memorial Lecture' given at the Smithsonian Institution in Washington, DC (8 September 2016). The chapter is a tribute to my Yupik partners from St. Lawrence, Alaska, particularly to Conrad Oozeva (1925–2016), Leonard Apangalook, Sr. (1939–2012), Chester Noongwook (1933–2020) and Willis Walunga (1925–2017). They were my mentors in research and, more than anyone else, have inspired my interest in the co-production of knowledge and in Indigenous observations and interpretations of Arctic change.

Note

1 My interpretation of the term 'memory-scape' is different here from the original meaning introduced by Mark Nuttall (1991).

References

Anonymous. 2019. Arctic sea ice reaches second lowest minimum in satellite record. National Snow and Ice Data Center, September 23, 2019 https://nsidc.org/arcticseai cenews/2019/09/arctic-sea-ice-reaches-second-lowest-minimum-in-satellite-record/ (accessed May 23, 2020)

Aporta, C. 2003. *Old Routes, New Trails: Contemporary Inuit Travel and Orienting in Igloolik, Nunavut*. PhD Thesis, Edmonton: University of Alberta.

Aporta, C. 2009. The trail as home: Inuit and their pan-Arctic network of routes. *Human Ecology*, 37: 131–146. https://DOI.org/10.1007/s10745–009-9213-x

Aporta, C. 2011. Shifting perspectives on shifting ice: Documenting and representing Inuit use of the sea ice. *The Canadian Geographer/ Le Géographe Canadien*, 55(1): 6–19. https://doi.org/10.1111/j.1541-0064.2010.00340.x

Arctic and Antarctic Research Institute (AARI). N.d. Ice terms arranged by subject. www .aari.ru/gdsidb/glossary/p1.htm (accessed May 23, 2020)

Bailey, A. M. 1943. The birds of Cape Prince of Wales, Alaska. *Proceedings of the Colorado Museum of Natural History*, 18(1).

Bank, T. P., II. 1953. Botanical and ethnobotanical studies in the Aleutian Islands II: Health and medical lore of the Aleuts. *Papers of the Michigan Academy of Science, Arts and Letters*, 38: 415–431.

Berlin, B., Breedlove, D. E. and Raven, P. H. 1966. Folk taxonomies and biological classification. *Science*, 154: 273–275.

Bogoslovskaya, L. S. and Krupnik, I. (eds.) 2013. *Nashi l'dy, snega i vetry. Narodnye i nauchnye znaniia o ledovykh landshaftakh i climate Vostochnoi Chukotki* [Our Ice, Snow, and Winds. Indigenous and Academic Knowledge on Ice-Scapes and Climate of Eastern Chukotka]. Moscow and Washington: Russian Heritage Institute.

Bulmer, R. 1957. A primitive ornithology. *Australian Museum Magazine*, 12(7): 224–229

Callaway, D. (with Eamer, J., Edwardsen, E., Jack, C., Marcy, S., Olrun, A., Patkoak, M., Rexford, D. and Whiting, A.). 1999. Effects of climate change on subsistence communities in Alaska. In Weller, G., and Anderson, P. A. (eds.) *Assessing the Consequences of Climate Change for Alaska and the Bering Sea Region: Proceedings of a Workshop (29–30 October 1998)*. Fairbanks: University of Alaska Fairbanks, pp. 59–74.

Conklin, H. C. 1955. 1955 Hanuóo color categories. *SouthWestern Journal of Anthropology*, 11: 339–344.

Dobrieva, E. A. and Alpergen, B. I. 2013. Ledovyi slovar' morskikh okhotnikov sela Naukan (Ice dictionary of sea-mammal hunters from the community of Naukan). In Bogoslovskaya, L. S. and Krupnik, I. (eds.) *Nashi l'dy, snega i vetry. Narodnye i nauchnye znaniia o ledovykh landshaftakh i climate Vostochnoi Chukotki* [Our Ice, Snow, and Winds. Indigenous and Academic Knowledge on Ice-Scapes and Climate of Eastern Chukotka]. Moscow and Washington: Russian Heritage Institute, pp. 148–153.

Druckenmiller, M. L., Daniel, R. and Johnson, M. (eds.) 2019. Voices from the front lines of a changing Bering Sea. An Indigenous perspective for the 2019 Arctic Report Card. In Richter-Menge, J., Druckenmiller, M. L. and Jeffries, M. (eds.). *Arctic Report Card 2019*, pp. 88–94. www.arctic.noaa.gov/Report-Card

Erdmann, F. 1864–1866. *Eskimoisches Wörterbuch gesammelt von Missionaren in Labrador, Revidirt und Herausgegeben von Friedrich Erdmann* [Eskimo Dictionary Collected by Missionaries in Labrador, Reviewed and Edited by Friedrich Erdmann]. Budissin [Bautzen], Germany: Ernst Moritz Monse, 2 vols.

Fienup-Riordan, A. and Rearden, A. 2012. *Ellavut / Our Yup'ik World and Weather. Continuity and Change on the Bering Sea Coast*. Seattle: University of Washington Press.

Fox (Gearheard), S. 2003. *When the Weather is Uggianaqtuq: Inuit Observations of Environmental Change*. Boulder: University of Colorado Geography Department Cartography Lab. Distributed by National Snow and Ice Data Center. CD-ROM.

Fox Gearheard, S., Holm, L. K., Huntington, H., Leavitt, J. M., Mahoney, A. R., Opie, M., Oshima, T. and Sanguya, J. (eds.) 2013. *The Meaning of Ice. People and Sea Ice in Three Arctic Communities*. Hanover, NH: International Polar Institute Press.

Freeman, M. M. R. 1984. Contemporary Inuit exploitation of the sea-ice environment. In Cooke, A. and Van Alstine, Edie (eds.) *Sikumiut: "The People Who Use the Sea Ice"*. Montréal: Canadian Arctic Resources Committee, pp. 73–96.

Gearheard, S., Matumeak, W., Angutikjuaq, I., Maslanik, J., Huntington, H. P., Leavitt, J., Matumeak-Kagak, D., Tigullaraq, G. and Barry, R. G. 2006. "It's not that simple": A comparison of sea ice environments, uses of sea ice, and vulnerability to change in Barrow, Alaska, USA and Clyde River, Nunavut, Canada. *AMBIO*, 35(4): 203–211. https://doi.org/10.1579/0044-7447(2006)35[203:INTSAC]2.0.CO;2

George, J. C. C., Huntington, H. P., Brewster, K., Eicken, H., Norton, D. W. and Glenn, R. 2004. Observations on shore-fast ice dynamics in Arctic Alaska and the responses of the Iñupiat hunting community. *Arctic*, 57(4): 363–374. https://doi.org/10.14430/arctic514

Hegger, D., Lamers, M., Van Zeijl-Rozema, A. and Dieperink, C. 2012. Conceptualising joint knowledge production in regional climate change adaptation projects: Success conditions and levers for action. *Environmental Science and Policy*, 18: 52–65

Heyes, S. A. 2011. Cracks in the knowledge: Sea ice terms in Kangiksualujjuaq, Nunavik. *The Canadian Geographer/Le Géographe canadien*, 55(1): 69–90. https://doi.org/10.1111/j.1541-0064.2010.00346.x

Höhn, O. E. 1969. Eskimo bird names at Chesterfield Inlet and Baker Lake, Keewatin, Northwest Territories. *Arctic*, 22(1): 72–75. https://doi.org/10.14430/arctic3195

Hovelsrud, G. K., Krupnik, I. and White, J. 2011. Human-based observing systems. In Krupnik, I., Allison, I., Bell, R., Cutler, P., Hik, D., Lopez-Martinez, J., Rachold, V., Sarukhanian, E. and Summerhayes, C. (eds.) *Understanding Earth's Polar Challenges. International Polar Year 2007-2008*. Edmonton: Canadian Circumpolar Institute, pp. 435–456.

Huntington, H. (ed.) 2000. Impacts of changes in sea ice and other environmental parameters in the Arctic. *Report of the Marine Mammal Commission Workshop*, Girdwood, Alaska, 15–17 February 2000. Bethesda: Marine Mammal Commission.

Huntington, H. P., Brower, Jr., H. and Norton, D. W. 2001. The Barrow symposium on sea ice, 2000: Evaluation of one means of exchanging information between subsistence whalers and scientists. *Arctic*, 54(2): 201–206 https://doi.org/10.14430/arctic780

Huntington, H. P. and Fox, S. 2005. The changing Arctic: Indigenous perspectives. In Symon, C., Arris, L. and Heal B. (eds.) *Arctic Climate Impact Assessment* (ACIA), New York: Cambridge University Press, pp. 61–98.

Irving, L. 1953. The naming of birds by Nunamiut Eskimo. *Arctic*, 6(1): 35–43. https://doi.org/10.14430/arctic3864

Irving, L. 1958. On the naming of birds by Eskimos. *Anthropological Papers of the University of Alaska*, 16(2): 61–77.

Jacobson, S. A. (ed.) 2008. *St. Lawrence Island-Siberian Yupik Eskimo Dictionary*, compiled by Womkom Badten, L., Oovi Kaneshiro, V., Oovi, M. and Koonooka, C. Fairbanks: Alaska Native Language Center.

Johns, A. 2010. On the Inuit sea ice terminology in Nunavut and Nunatsiavut. In Krupnik, I., Aporta, C., Gearheard, S., Laidler, G. J. and Kielsen Holm, L. (eds.) *SIKU: Knowing Our Ice. Documenting Inuit Sea Ice Knowledge and Use*. Dordrecht: Springer, pp. 401–412. https://doi.org/10.1007/978-90-481-8587-0_17

Johnson, N., Behe, C., Danielsen, F., Krümmel, E-M., Nickels, S. and Pulsifer, P. L. 2016. *Community-Based Monitoring and Indigenous Knowledge in a Changing Arctic: A Review for the Sustaining Arctic Observing Networks*. Ottawa: Inuit Circumpolar Council Canada www.inuitcircumpolar.com/uploads/3/0/5/4/30542564/cbm_final_report.pdf

Jolly, D., Berkes, F., Castleden, J., Nichols, T. and the community of Sachs Harbour. 2002. "We can't predict the weather like we used to": Inuvialuit observations of climate Change, Sachs Harbour, Western Canadian Arctic. In Krupnik, I., and Jolly, D. (eds.) *The Earth Is Faster Now: Indigenous Observations of Arctic Environmental Change*. Fairbanks, AK: ARCUS, pp. 92–125.

Kleinschmidt, S. 1871. *Den grønlandske Ordbog* [The Greenlandic Dictionary]. Copehagen: Louis Kleins Bogtrykkeri.

Krupnik, I. 2000. Native perspectives on climate and sea ice changes. In Huntington, H. P. (ed.) *Impacts of Changes in Sea Ice and Other Environmental Parameters in the Arctic. Report of the Marine Mammal Commission Workshop*, Girdwood, Alaska, 15–17 February 2000. Bethesda: Marine Mammal Commission, pp. 25–39.

Krupnik, I. 2002. Watching ice and weather our way: Some lessons from Yupik observations of sea ice and weather on St. Lawrence Island, Alaska. In Krupnik, I. and Jolly, D. (eds.) *The Earth Is Faster Now: Indigenous Observations of Arctic Environmental Change*. Fairbanks, AK: ARCUS, pp. 156–197.

Krupnik, I. 2011. 'How many Eskimo words for ice?' Collecting Inuit sea ice terminologies in the International Polar Year 2007–2008. *The Canadian Geographer/Le Géographe canadien*, 55(1): 56–64. https://doi.org/10.1111/j.1541-0064.2010.00345.x

Krupnik, I. and Jolly, D. 2002. (eds.) *The Earth Is Faster Now: Indigenous Observations of Arctic Environmental Change*. Fairbanks, AK: ARCUS (2nd edition, 2010).

Krupnik, I. and Huntington, H. P. 2004. Introduction. In Oozeva, C., Noongwook, C., Noongwook, G., Alowa, C. and Krupnik, I. *Sikumengllu Eslamengllu Esghapaleghput /Watching Ice and Weather Our Way*. Washington, DC: Arctic Studies Center, pp.16–21

Krupnik, I. and Chlenov, M. 2007. The end of "Eskimo Land": Yupik relocations, 1958–1959. *Etudes/Inuit/Studies*, 31(1–2): 59–82. https://doi.org/10.7202/019715arCopied

Krupnik, I. and Müller-Wille, L. 2010. Franz Boas and Inuktitut terminology for ice and snow: From the emergence of the field to the "Great Eskimo vocabulary hoax". In Krupnik, I., Aporta, C., Gearheard, S., Laidler, G. J. and Kielsen Holm, L. (eds.) *SIKU: Knowing Our Ice. Documenting Inuit Sea Ice Knowledge and Use*. Dordrecht: Springer, pp. 377–400. https://doi.org/10.1007/978-90-481-8587-0_16

Krupnik, I. and Weyapuk, Jr., W. 2010. *Qanuq Ilitaavut* : "How we learned what we know" (Wales Inupiaq Sea Ice Dictionary). In Krupnik, I., Aporta, C., Gearheard, S., Laidler, G. J. and Kielsen Holm, L. (eds.) *SIKU: Knowing Our Ice. Documenting Inuit Sea Ice Knowledge and Use*. Dordrecht: Springer, pp. 321–354. https://doi.org/10.1007/978-90-481-8587-0_14

Krupnik, I. and Bogoslovskaya, L. S. 2017. "Our ice, snow and winds": From knowledge integration to co-production in the Russian SIKU project, 2007–2013. In Kasten, E., Roller, K. and Wilbur, J. (eds.) *Oral History Meets Linguistics*. Halle: Verlag der Kulturstiftung Sibirien, pp. 31–48.

Krupnik, I., Aporta, C., Gearheard, S., Laidler, G. J. and Holm, L. K. (eds.) 2010a. *SIKU: Knowing Our Ice. Documenting Inuit Sea Ice Knowledge and Use*. Dordrecht: Springer.

Krupnik, I., Aporta, C. and Laidler, G. J. 2010b. SIKU: International Polar Year Project #166 (An Overview). In Krupnik, I., Aporta, C., Gearheard, S., Laidler, G. J. and Kielsen Holm, L. (eds.) *SIKU: Knowing Our Ice. Documenting Inuit Sea Ice Knowledge and Use*. Dordrecht: Springer, pp. 1–28.

Krupnik, I., Apangalook, L. and Apangalook, P. 2010c. "It's cold but not cold enough": Observing ice and climate change in Gambell, Alaska in IPY 2007–2008 and beyond. In Krupnik, I., Aporta, C., Gearheard, S., Laidler, G. J. and Kielsen Holm, L. (eds.) *SIKU: Knowing Our Ice. Documenting Inuit Sea Ice Knowledge and Use*. Dordrecht: Springer, pp. 81–114. https://doi.org/10.1007/978-90-481-8587-0_4

Laidler, G. J. 2006. Inuit and scientific perspectives on the relationship between sea ice and climate change: The ideal complement? *Climatic Change*, 78(2–4): 407–444. https://doi.org/10.1007/s10584-006-9064-z

Laidler, G. J. 2007. *Ice, Through Inuit Eyes: Characterizing the Importance of Sea Ice Processes, Use and Change around Three Nunavut Communities*. PhD Thesis, Toronto, ON: University of Toronto.

Laidler, G. J., and Elee, P. 2008. Human geographies of sea ice: Freeze/thaw processes around Cape Dorset, Nunavut, Canada. *Polar Record*, 44: 51–76

Laidler, G. J., and Ikummaq, T. 2008. Human geographies of sea ice: Freeze/thaw processes around Igloolik, Nunavut, Canada. *Polar Record*, 44: 127–153

Laidler, G. J., Elee, P., Ikummaq, T., Joamie, E. and Aporta, C. 2010. Mapping Inuit sea ice knowledge, use and change in Nunavut, Canada. In Krupnik, I., Aporta, C., Gearheard, S., Laidler, G. J. and Kielsen Holm, L. (eds.) *SIKU: Knowing Our Ice. Documenting Inuit Sea Ice Knowledge and Use*. Dordrecht: Springer, pp. 45–80 https://doi.org/10.1007/978-90-481-8587-0_3

Lantis, M. 1959. Folk medicine and hygiene: Lower Kuskokwim and Nunivak-Nelson Island areas. *Anthropological Papers of the University of Alaska*, 8(1): 1–75.

Larsen, J. N. and Anisimov, O. A. 2014. Climate Change 2014: Impact, Adaptation, and Vulnerability. Part B – Regional Aspects: Polar Regions. In Barros, V. R. and Field, C. B. (eds.). *Working Group II Contribution to the Fifth Assessment Report of the Intergovernmental Panel on Climate Change*. Cambridge: Cambridge University Press, pp. 1567–1612.

Leonova, V. G. 2014. *Naukan i naukantsy* [Naukan and the Naukanese]. Vladivostok: Dalpress.

Lindsey, R. and Scott, M. 2019. Climate Change: Arctic sea ice summer minimum. Climate.gov. September 26, 2019 – www.climate.gov/news-features/understanding-climate/climate-change-minimum-arctic-sea-ice-extent (accessed May 20, 2020)

MacLean, E., comp. 2014. *Iñupiatun Uqaluit Taniktun Sivuninit/Iñupiaq to English Dictionary*. Fairbanks: University of Alaska Press.

McDonald, M., Arragutainaq, L., and Novalinga, Z. 1997. *Voices from the Bay: Traditional Ecological Knowledge of Inuit and Cree in the Hudson Bay Bioregion*. Ottawa, Sanikiluaq: Canadian Arctic Resources Committee, Municipality of Sanikiluaq.

Macpherson A. H. 1958. Arviligjuarmiut names for birds and mammals. *Arctic Circular*, 9: 30–34

Morgounova Schwalbe, D. 2015. Language ideologies at work: Economies of Yupik language maintenance and loss. *Sibirica*, 14(3): 1–27.

Nakashima, D. 1993. Astute observers on the sea ice edge: Inuit knowledge as a basis for Arctic co-management. In Inglis, J. T. (ed.) *Traditional Ecological Knowledge:*

Concepts and Cases. Ottawa: International Programme on Traditional ecological Knowledge.

Nelson, R. K. 1969. *Hunters of the Northern Ice*. Chicago: University of Chicago Press.

Nichols, T., Berkes, F., Jolly, D., Snow, N. B. and the Community of Sachs Harbour. 2004. Climate change and sea ice: Local observations from the Canadian Western Arctic. *Arctic*, 57(1): 68–79.

Nickels, S., Furgal, C., Buell, M. and Moquin, H. 2005. *Unikkaaqatigiit: Putting the Human Face on Climate Change: Perspectives from Inuit in Canada*. Ottawa: Inuit Tapiriit Kanatami, Nasivvik Centre for Inuit Health and Changing Environments at Université Laval, and the Ajunnginiq Centre at the National Aboriginal Health Organization.

Norton, D. W. 2002. Coastal sea ice watch: Private confessions of a convert to Indigenous knowledge. In Krupnik I. and Jolly, D. (eds.) *The Earth Is Faster Now: Indigenous Observations of Arctic Environmental Change*. Fairbanks, AK: ARCUS, pp.126–155.

Nutawyi, A. and Radunovich, N. 2013. Ledovyi slovar' morskikh okhotnikov sela Sireniki (na yazyke chaplinski yupik) (Ice Dictionary of Maritime Hunters in the Community of Sireniki (Chaplinski Yupik language). In Bogoslovskaya, L. S. and Krupnik, K. (eds.) *Nashi l'dy, snega i vetry. Narodnye i nauchnye znaniia o ledovykh land-shaftakh i climate Vostochnoi Chukotki*. Moscow and Washington: Russian Heritage Institute, pp. 72–82.

Nuttall, M. 1991. Memoryscape: A sense of locality in Northwest Greenland. *North Atlantic Studies*, 1(2): 39–50.

Oozeva, C., Noongwook, C., Noongwook, G., Alowa, C. and Krupnik, I. 2004. *Sikumengllu Eslamengllu Esghapaleghput /Watching Ice and Weather Our Way*. Washington, DC: Arctic Studies Center.

Oswalt, W. H. 1957. A Western Eskimo Ethnobotany. *Anthropological Papers of the University of Alaska*, 6(1): 16–36.

Perovich, D., Meier, W., Tschudi, M., Farrell, S., Hendricks, S., Gerland, S., Haas, C., Krumpen, T., Polashenski, C., Ricker, R. and Webster, W. 2018. Sea Ice. pp. 25-32 in *Arctic Report Card 2018* www.arctic.noaa.gov/Report-Card (accessed January 3, 2020)

Petitot, É. P. 1876. *Vocabulaire Français-Esquimau : Dialecte des Tchiglit dès bouches du Mackenzie et de l'Anderson*. Paris: Ernest Leroux.

PIOMAS. 2016. Arctic sea ice volume reanalysis. http://psc.apl.uw.edu/research/projects/arctic-sea-ice-volume-anomaly/ (accessed September 17, 2016)

Pohl C., Rist, S., Zimmermann, A., Fry, P., Gurung, G., Schneider, F., Speranza, C. I., Kiteme, B., Boillar, S., Serrano, E., Hirsch Hadorn, G. and Wiesmann, U. 2010. Researchers' role in knowledge co-production: Experience from sustainability research in Kenya, Switzerland, Bolivia, and Nepal. *Science and Public Policy*, 37 (4): 267–281.

Pungowiyi, C. (2000). Native observations of change in the marine environment of the Bering Strait region. In *Impacts of Changes in Sea Ice and Other Environmental Parameters in the Arctic*. Final Report of the Marine Mammal Commission Workshop, Girdwood, Alaska, 15–17 February 2000.

Ray, C. and Hufford, G. 1989. Relationships among Beringian marine mammals and sea ice. *Rapports et Procès-Verbaux des Réunions (Oceanography and Biology of Arctic Seas)*, 188: 22–39

Ray, C. G., Hufford, G. L., Overland J. O., Krupnik I., McCormick R., Jerry, Frey, Karen, and Labunski, E. 2016. Decadal Bering Sea seascape change: Consequences for Pacific walruses and Indigenous hunters. *Ecological Applications*, 26(1): 24–41.

Riedlinger, D. and Berkes, F. 2001. Contributions of traditional knowledge to understanding climate change in the Canadian Arctic. *Polar Record*, 37(203): 315–328.

Riewe, R. 1991. Inuit use of the sea ice. *Arctic and Alpine Research*, 23(1): 3–10.

Salomon, A., Huntington, H. P. and Tanape, Sr., N. 2011. *Imam cimucia: Our Changing Sea*. Fairbanks, AK: Alaska Sea Grant.

Sturtevant, W. C. 1964. Studies in ethnoscience. *American Anthropologist*, 66(3) pt. 2: 99–131.

Stegmaier, P. 2009. The rock 'n' roll of knowledge co-production. *EMBO Reports*, 10(2): 114–119.

Taverniers, P. 2010. Weather variability and changing sea ice use in Qeqertaq, West Greenland, 1987–2008. In Krupnik, I., Aporta, C., Gearheard, S., Laidler, G. J. and Kielsen Holm, L. (eds.) *SIKU: Knowing Our Ice. Documenting Inuit Sea Ice Knowledge and Use*. Dordrecht: Springer, pp. 31–44. https://doi.org/10.1007/978-90-481-8587-0_2

Tersis, N. and Taverniers, P. 2010. Two Greenlandic sea ice lists and some considerations regarding Inuit sea ice terms. In Krupnik, I., Aporta, C., Gearheard, S., Laidler, G. J. and Kielsen Holm, L. (eds.) *SIKU: Knowing Our Ice. Documenting Inuit Sea Ice Knowledge and Use*. Dordrecht: Springer, pp. 413–426. https://doi.org/10.1007/978-90-481-8587-0_18

Vaughan, D. G., Comiso, J. C., Allison, I., Carrasco, J., Kaser, G., Kwok, R., Mote, P., Murray, T., Paul, F., Ren, J., Rignot, E., Solomina, O., Steffen K. and Zhang, T. 2013. Observations: Cryosphere. In Stocker, T. F., Qin, D., Plattner, G.-K., Tignor, M., Allen, S. K., Boschung, J., Nauels, A., Xia, Y., Bex V. and Midgley, P. M. (eds.) *Climate Change 2013: The Physical Science Basis*. Cambridge and New York: Cambridge University Press.

Walunga, W., comp. 1986. *St. Lawrence Island Curriculum Resource Manual*. Gambell and Unalakleet, AK: Bering Strait School District.

Weyapuk, Jr., W. and Krupnik, I. 2012. *Kiŋikmi Sigum Qanuq Ilitaavut/ Wales Inupiaq Sea Ice Dictionary*. Washington: Arctic Studies Center and the Native Village of Wales.

Wisniewski, J. 2010. Knowings about sigu: Kigiqtaamiut hunting as an experiential pedagogy. In Krupnik, I., Aporta, C., Gearheard, S., Laidler, G. J. and Kielsen Holm, L. (eds.) *SIKU: Knowing Our Ice. Documenting Inuit Sea Ice Knowledge and Use*. Dordrecht: Springer, pp. 275–294. https://doi.org/10.1007/978-90-481-8587-0_12

Yashchenko, O. 2016. "Smotri i uchis": molodye chukotskie i eskimosskie morskie okhotniki sel Lorino i Sireniki 2010-kh gg. ("Look and Follow Me": Young cohort of Chukchi and Yupik maritime hunters in the communities of Lorino and Sireniki in the 2010s). In Krupnik, I. (ed.) *Litsom k moriu. Pamiati Lyudmily Bogoslovskoi*. Moscow: August Borg, pp. 191–213.

Young, S. B., and Hall, Jr., E. S. 1969. Contributions to the ethnobotany of the St. Lawrence Island Eskimo. *Anthropological Papers of the University of Alaska*, 14(2): 43–53

6

Mapping Land Use with Sámi Reindeer Herders: Co-production in an Era of Climate Change

MARIE ROUÉ, LARS-EVERT NUTTI, NILS-JOHAN UTSI
AND SAMUEL ROTURIER

This chapter presents research co-produced in 2009 between interdisciplinary scientists and reindeer herders from the Sámi community of Sirges in northern Sweden. The research project was conceived in response to a participatory mapping programme known as the Reindeer Husbandry Plan (RHP), led by the Swedish Forest Agency. Financed from the year 2000 to 2014, the RHP uses a digital programme to compile and map habitat use by reindeer herding communities (*sameby* in Swedish). It is a management tool that, according to the Swedish Forest Agency (2003), contributes to the resolution of land use conflicts by, 'facilitate[ing] the dialogue [of reindeer herders] with other land users, such as the timber industry'.

However, as initially conceived and implemented, the RHP applied preconceived categories and standardized methods to identify pastures assumed to be the 'best' for reindeer husbandry. This approach raised a number of serious concerns for the authors, ecologist Samuel Roturier and anthropologist Marie Roué. Based on their previous research with reindeer herders, they knew that impermanence and adaptability in the face of highly variable pasture conditions are at the heart of Sámi land use. The availability of forage in winter is of critical importance, in particular, access to lichen under snow cover. For this reason, Sámi herd management is adaptive and based on a continuous monitoring of snow and weather conditions. The fixed and hierarchical land categories established by the RHP do not take this inherent variability of pasture conditions into account. Given the already tense relations between the forestry industry and reindeer herders, such a management tool, despite initial good intentions, ran the risk of engendering grave misunderstandings.

Roué and Roturier proposed the organization of a 'co-production workshop' during which the Sámi could reconsider the RHP in the context of the complex reality of their pasture use. The workshop would provide an opportunity to reflect on the strengths and weaknesses of the initial RHP methodology, while exploring avenues to improve its implementation and effectiveness through Sámi knowledge

117

of the complex relationships among weather conditions, land use and the well-being of their herds. While a workshop by itself may not resolve a major resource use conflict rooted in an unequal access to power, the co-production of knowledge might, nonetheless, provide a more solid foundation for co-management and a degree of empowerment for the Sámi. Before describing the methods and outcomes of this exploratory workshop, the RHP as initially conceived and put into practice is briefly presented.

The Reindeer Husbandry Plan: A Participatory Plan for Reindeer Herders

The Reindeer Husbandry Plan (RHP), commonly known in Sweden as the *Renbruksplan*, is a digital tool 'for communication and land use planning' by reindeer herders and reindeer herding communities. It was modelled after the 'forest management plan' (*Skogsbruksplan)*, a decision-making tool 'designed by and for forest owners to assess and manage their forests'. The RHP was expected to 'enhance relations between reindeer herders and forest managers, especially those from large forestry companies' (Swedish Forest Agency, 2001; Sandström et al., 2003). According to the Swedish Forest Agency (2014), the RHP traces its origins to a herder who remarked to his daughter that herders should design their own plans. She then introduced this idea to the Västerbotten county administration, where she was employed. They took up the proposal in collaboration with various local and national partners, including the Swedish Forest Agency. While this account emphasizes the participatory element, the RHP also owes its origins to an omnipresent culture of land use planning.

From the beginning, the project was financed primarily by the Swedish government and directed by the Swedish Forest Agency. The RHP funded each reindeer herding community to develop their local RHP by mapping grazing areas in a custom-made geographic information system (GIS) named RenGIS. Conceived as a participatory GIS, herders receive basic training in GIS, field inventory techniques and satellite-image interpretation in order to map important areas such as calving grounds and pastures during different seasons. The RHP organized visits by reindeer herders to their winter grazing areas during the snow-free season, in order to estimate lichen cover and biomass of terrestrial and arboreal lichens.

At the heart of the RHP is the classification and digital mapping of areas of importance, using a computer screen with a satellite image of the territory displayed in the background (Swedish Forest Agency, 2003). The satellite image allowed herders, instructed by GIS planners, to discriminate fairly accurately between different types of vegetation. Sámi herders learned very quickly to recognize lichen-rich pine forests on the images (Fig. 6.1).

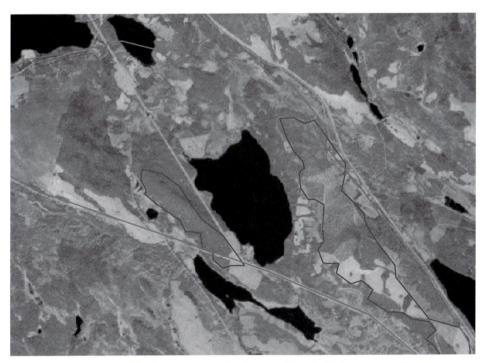

Figure 6.1 The satellite image used as background for digitizing core and key areas for the RHP. The Red/Green/Blue combination display allowed fairly accurate discrimination of forest vegetation. The lichen-rich pine forests are easily identifiable by their blue-grey colour. One core area (blue lines) and two key areas (red lines) are shown on this image. [A black and white version of this figure will appear in some formats. For the colour version, please refer to the plate section.]
Source: Sirges RHP

Based on this training, and by mobilizing their knowledge, herders were invited to classify their land into three inclusive and hierarchical categories: (1) 'the general seasonal grazing area', which is the largest and most inclusive; (2) the 'core area'; and (3) the 'key area', the most restricted, and in theory, the most important (see Table 6.1 for definitions). These categories are clearly derived from a protected area model. The term 'core area' is derived from the 1970s Man and the Biosphere (MAB) programme that establishes biosphere reserves (see Batisse, 1986), and the term 'key area' relates to the 'key biotope', a category widely used in forestry and nature conservation in Sweden.

In theory, herders had full freedom to designate their key, core and seasonal grazing areas. In practice, however, the methodology closely adhered to ecological definitions of habitat based on vegetation and natural resource management principles. Furthermore, as shown in Table 6.1, the three categories are not clearly differentiated and have limited relevance for the land use practices of the herders.

Table 6.1 *Definitions of the categories used to map the RHP*

Category	Swedish term	Definition translated from original Swedish (from Swedish Forest Agency, 2003)
General seasonal grazing area	*Betestrakt*	Land surrounding a core area. Grazing area where reindeer are kept. Size varies depending on the season. Several 'general seasonal grazing areas' can overlap.
Core area	*Kärnområde*	Important areas regularly used in reindeer husbandry. Core areas often include several important grazing areas where disturbance must be minimized for grazing animals. May include several key areas that are too small to be mapped. Sensitive to encroachment.
Key area	*Nyckelområde*	The most important areas, often islands within core areas, that reindeer naturally frequent. Very sensitive to encroachment and disturbance.

It is important to note that herders did not participate in the development of RHP methodologies and their traditional knowledge was not solicited for the establishment of categories or any other aspect of the plan.

The RHP was implemented in three phases: a pilot study with two Sámi communities (2000–2002); a second phase extending the programme to four volunteer communities, which included the community of Sirges (2005–2010); and a final phase during which all fifty-one communities had to complete their RHP (2011–2014). The co-production workshop took place in 2009, towards the end of the RHP's second phase. Although it was not officially part of the RHP process, the timing of the workshop was such that it contributed to a reflection on phases one and two of the RHP. By questioning initial RHP modalities and introducing novel methods that incorporate Sámi experience and knowledge, this co-produced workshop had an impact on subsequent RHP implementation, as well as on its final outcome.

The Challenge of Mapping Impermanence and Complexity

From 2007 to 2009, Roué and Roturier conducted an interdisciplinary study on Sámi ecological knowledge of the variability of winter pasture conditions. Their results demonstrated that the Sámi possess a sophisticated science of the interaction between snow and pastures, including a discriminating knowledge of extreme weather events and rapid changes in snow and ice conditions (Roturier and Roué, 2009). This interdisciplinary academic research entered a new phase with the opportunity in 2009 to engage in a dialogue with both Sámi and the RHP. The workshop, conducted in agreement and in collaboration with members of both

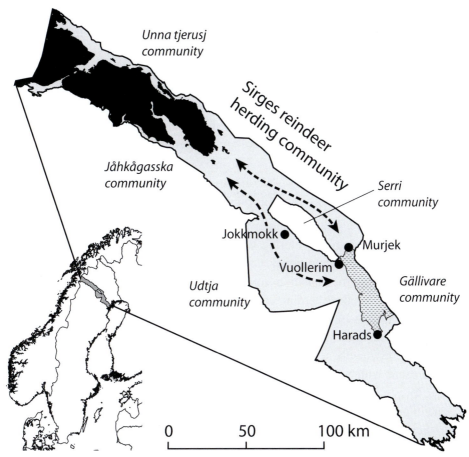

Figure 6.2 The Sirges reindeer herding community in northern Sweden. The reindeer herds migrate on a seasonal basis between summer pastures in alpine tundra and birch forests to the west (dark area), down to winter pastures in lichen-rich conifer forests to the east. The hatched area represents the winter territory of the two winter groups, or *siidat*, who contributed to the co-production workshop. Map by S. Roturier

the Sirges reindeer herding community (Fig. 6.2) and the RHP, led to a co-production of knowledge directed towards the needs of the herders, who were themselves active co-researchers in the undertaking.

The three-day workshop was organized with two winter groups, or *siidat* (in Sámi), of reindeer herders from the Sirges community. The goal was to discuss the RHP and eventually refine the husbandry plans developed for their community. Per Sandström from the Swedish University of Agricultural Sciences, responsible for the RHP, was contacted and he expressed his interest in participating in this exploratory workshop. The objective was to better understand Sámi usage of pastures and their adaptation strategies in the face of environmental variability, including climate change. In other words, the authors proposed to elaborate a Sámi view of husbandry planning. To this end, it was decided to concentrate on winter, the most challenging season for reindeer husbandry, when herders must fully deploy their adaptive capacity. This mapping went far beyond the initial participatory mapping of the RHP. Its innovative methodology brought together Sámi knowledge and land use, environmental anthropology, forest ecology and geomatics.

A Methodology for Mapping Sámi Land Use in Winter

A 'winter-group' is a group of herder families that constitute a sub-unit within the reindeer herding community during winter. In Sámi, this sub-unit is known as a *siida* (pl. *siidat*), the traditional cooperative social unit that existed several centuries ago before the transition from a hunting to a herding economy. The entity designated as a 'reindeer herding community' (*sameby* in Swedish, also translated as 'Sámi village' in English) was created in 1886 by the Swedish Crown. Both a geographic unit delimiting grazing areas and an economic organization representing its members' interests, it encapsulates several *siidat* and covers the entire area in which the herders and their reindeer live and migrate, from the conifer forests used in the winter that extend down to the Baltic sea to the high mountain pastures used in summer (Fig. 6.2). Today, the *sameby* is the only social unit taken into consideration by the administration. Traditionally, however, there was no single entity that existed throughout the year. Rather, Sámi herders organized themselves into smaller winter *siidat* and more inclusive summer *siidat*. The term 'reindeer herding community', like the term *siida*, also refers to a community in the more commonly understood sense, that is, the families who share the land and together undertake many essential activities. While the RHP adopted the *sameby* or reindeer herding community as its base unit, our co-production workshop mapped winter land use at the level at which decisions and strategies are elaborated, that is, the winter *siida*.

To understand the pasture management and land use patterns of the two winter *siida*, herders mapped the displacement of their herds throughout each winter and provided information on relevant aspects of reindeer husbandry.

- When did they move their herds from one place to another?
- What was the reasoning behind their decisions and strategies?
- What were the weather and pasture conditions like?
- How did they respond when an ice crust developed on top or within the snowpack, blocking access to the lichen?

Figure 6.3 Reindeer herders mapping their land use during the co-production workshop held in Vuollerim, Sweden, in February 2009.
Photos by M. Roué/S. Roturier

GIS maps served as a reference, data was recorded on large-format paper maps and additional information was captured on film, audiotape and via notetaking (Fig. 6.3).

The methodology of 'retrospective land use' was based on the systematic collection of data for current and past years. Herders first documented their land use during the (ongoing) winter of 2008–2009 and then tracked back over four previous winters. The advantage of this approach is that mapping begins with a familiar situation (the current winter) that herders record with high resolution in time and space: where are the reindeer now and where did they graze earlier this winter? In this way, salient issues are identified for the mapping of preceding winters. Tracing back four years was appropriate, as herder memories proved to be fully reliable over this period. When necessary, herders cross-checked their recollections against administrative documents (e.g., invoices), their own almanacs or photographs. In spring 2010, this data collection was repeated, thus adding the winter of 2009–2010 to the data set. Herders and researchers were thus able to establish a multi-year database comparing conditions and strategies across several years.

In contrast with the static hierarchy of pastures mapped by the RHP, herders consider a multitude of factors in order to adjust the pattern of pasture use and the timing of migration from one pasture to the next. Consequently, the methodology placed particular emphasis on herd displacements. This focus on movement rather than on permanence, based on actual land use rather than on general categories, contrasted sharply with the RHP.

Managing a herd of reindeer in a subarctic environment is, by necessity, a highly adaptive undertaking – a planning for impermanence. The Sámi have developed a sophisticated science of snow and pasture interaction (Magga, 2006; Roturier and Roué, 2009; Riseth et al., 2011; Eira et al., 2012), and a specialized vocabulary to describe, monitor and manage pastures (Ruong, 1964; Jernsletten, 1997; Ryd, 2007). During the co-production workshop, the elaboration of a taxonomy of Sámi terms for land use categories was valuable for understanding and facilitating cross-cultural knowledge exchange, thus avoiding cumbersome paraphrases for Sámi terms that lack Swedish equivalents. It also empowered the herders, placing them in the lead: a major change from the RHP where a Western worldview prevails through the use of the terms and categories of ecologists and managers.

The central role of Sámi concepts is illustrated by the term *guohtun*, glossed in English as 'pasture' (Roturier and Roué, 2009). Its meaning in Sámi, however, is more specific and dynamic: a pasture that reindeer can access and graze at a specific moment in time. Pasture quality is not an absolute. A place rich in lichens cannot simply be mapped as *guohtun* because it may cease to exist overnight. The formation of an ice crust will prevent reindeer from digging through the snow and grazing lichens, which, even though still abundant, suddenly become inaccessible. So, the main factor determining the quality of a pasture is not the absolute abundance of lichen, but the nature of the snow cover that at any given moment determines its accessibility.

Terms for critical land categories in a Sámi worldview were also instrumental for discussions. For a non-Sámi person, a landscape blanketed with a fresh layer of snow, as yet undisturbed by human or animal activity, may seem 'pristine'. Sámi herders, however, consider the history of the land's use and not the snow surface alone. The Sámi term *oppas* refers to pasture that has not previously been grazed by the herd. Here, the reindeer can find new and accessible lichen to graze. In opposition, *čiegar* is a snowfield that has already been grazed that winter by reindeer and is thus trampled and compacted. When they are both covered by a fresh layer of snow, they may look the same. But a *čiegar* is of no use because the underlying compacted snow prevents further grazing. Thus, throughout the winter the Sámi move their herds from a pristine *oppas* where they graze with ease, transforming it into a *čiegar*, before moving on to a new *oppas*.

Pasture Management in a Changing Climate: Workshop Results

The three-day co-production workshop in Vuollerim was located between the territories of the two winter-groups of the Sirges reindeer herding community, whose territories are adjacent to one another. The Sámi herders who participated in the workshop were Lars-Evert Nutti (born 1959), Per-Olof Kuhmunen (born 1946), Nils-Johan Utsi (born 1947) and his son Carl-Johan Utsi (born 1978). Lars-Evert Nutti and Nils-Johan Utsi, co-authors of this chapter, were the referent persons for the RHP in their respective *siida*. The winter-group of Lars-Evert Nutti and Per-Olof Kuhmunen uses a strip of land (about 70 x 25 km) east of the Lule river, north of Harads (Fig. 6.2). The *siida* Utsi grazing lands extend over a smaller area (about 10×40 km) south of Murjek.

The following descriptions summarize the patterns of land use and pasture management by these two *siida* over six consecutive winters from 2004–2005 to 2009–2010. They illustrate how the herders and their animals dealt with the variability of weather and grazing conditions and other environmental constraints.

Winter 2004–2005 (Fig. 6.4a) was considered by the herders to be a good winter, with no negative weather events to reduce the quality of grazing conditions.

Adaptive Strategies The *siida* Nutti restricted their reindeer grazing to limited areas to preserve other parts of the land. The *siida* Utsi grazed their territory in their habitual manner. Both *siida* were able to migrate by foot with their herds in spring.

Winter 2005–2006 (Fig. 6.4b) was described as a good winter. Temperature and snow conditions in 'winter-spring'[1] created good grazing conditions that continued late into the spring season. During this season, repeated thaw–freeze events, usually in April, result in the formation of a 'crust of ice on top of the snowpack', known as *skávvi*, which creates good conditions for moving the reindeer over long distances. This is the right moment to begin the migration to calving grounds.

Adaptive Strategies The *siida* Nutti started by grazing the lichen-rich pastures adjacent to the railway line. Grazing these areas early in winter is essential as, later in the season, accumulating snow buries the fence, allowing reindeer to stray onto the tracks where they may be killed by trains. During that winter, reindeer were kept grazing in a large, continuous lichen-rich area from which they didn't move until winter-spring. The herders left their reindeer grazing on the winter territory over a longer period than usual to take advantage of the good grazing conditions. They postponed their migration to the calving grounds until the end of April, only a couple of days before calving.

Winter 2006–2007 (Fig. 6.4c) was considered catastrophic with respect to grazing conditions. In November, a thaw–freeze degraded grazing conditions over

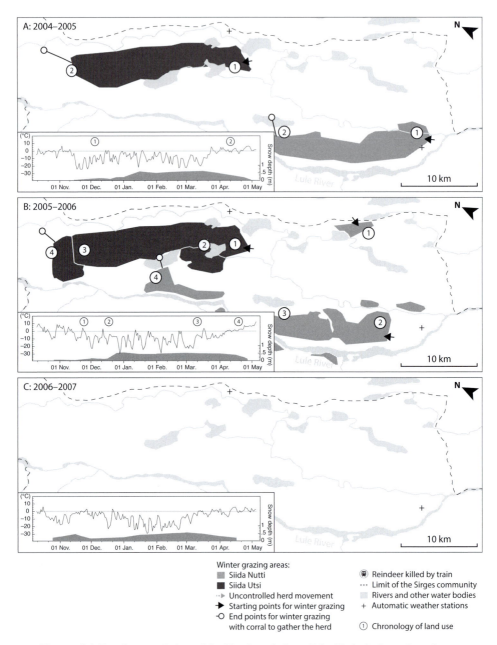

Figure 6.4 Land use of the *siida* Nutti and the *siida* Utsi during six winters between 2004 and 2010, including the chronology of herd locations throughout each winter. Snow depth and temperatures during the winter are presented in the insets. Thaw–freeze events may occur when temperatures rise above 0°C (horizontal line).

Maps by S. Roturier

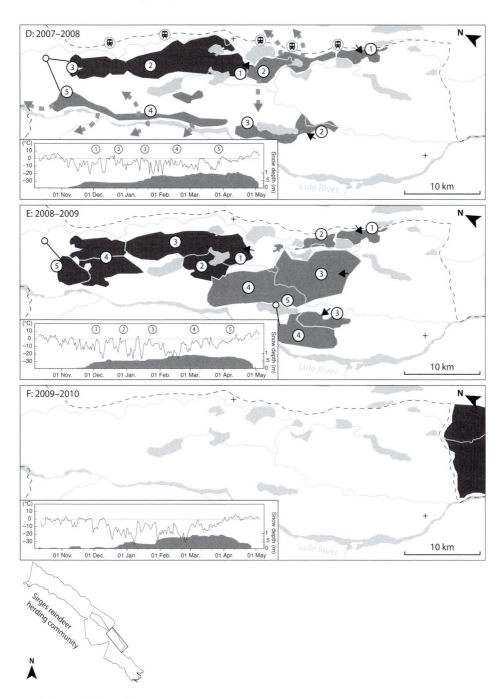

Figure 6.4 (*cont.*)

Figure 6.5 Artificial feeding of reindeer in late spring. Photo by S. Roturier. [A black and white version of this figure will appear in some formats. For the colour version, please refer to the plate section.]

a large area of northern Sweden, forming a crust of ice and hard snow on top of the vegetation that prevented reindeer from gaining access to ground lichens.

Adaptive Strategies Neither of the two *siida* could use the traditional wintering area for grazing during the winter of 2006–2007. Instead, part of their herds, namely the yearlings, were moved to other locations and fed pellets. Artificial feeding is extremely costly in terms of money and labour (Fig. 6.5). It is also risky, as many reindeer do not manage to adapt their stomach flora, causing death in some cases. The *siida* Nutti decided to overwinter a large part of their herd in the fall territory, the lowest part of their mountain pastures. As in many other Sámi communities in Sweden, they applied for economic compensation that covered about half the cost of the pellets.

Winter 2007–2008 (Fig. 6.4d) was characterized by adverse snow conditions, in particular an extreme thaw–freeze event during the last two weeks of December, to which the two *siida* responded in very different ways. The situation was particularly catastrophic for the *siida* Nutti: many of their reindeer were already grazing on the winter pastures when they brought the remainder of the herd from their fall territory. Immediately after, snow conditions rapidly degraded, resulting in a hard snow layer that reindeer could not penetrate. The reindeer responded by scattering rapidly in all directions in a desperate attempt to find better pastures.

Adaptive Strategies Almost immediately, in January, the *siida* Utsi gathered its herd into a corral and transported it by truck for release in the mountain area normally used only as a summer pasture. This highly unusual strategy was judged to be a success in the end. Their reindeer survived and produced many calves. In the meantime, the herders of the *siida* Nutti were in a desperate situation, having lost control of a herd that was scattering in all directions searching for accessible pastures. The herders struggled for many weeks, trying to gather the reindeer, while the main body of the herd kept moving rapidly northward because the grazing conditions remained very poor. Grazing had to be complemented by feeding with pellets. Furthermore, many reindeer were killed, struck by trains when crossing the railway tracks in search of better pastures. At the end of the season, only half of the reindeer were gathered in a corral at the north end of their winter territory before being transported to the calving grounds at the end of March. The herders had to abandon gathering the other half and could only hope that the reindeer would find their own way to the calving lands. During the following summer and autumn, the herders had the extra workload of trying to locate reindeer that had strayed into the territories of neighbouring *siida*.

Winter 2008–2009 (Fig. 6.4e) was described as a normal winter, with no particularly adverse snow conditions.

Adaptive Strategies Throughout the winter, the reindeer of the two *siida* kept moving north-west as they grazed. The herders' main task was to keep their herds from intermingling. In winter-spring, the reindeer were kept in the vicinity of the corrals, where they were finally gathered in mid-April to be moved by truck to the calving lands.

Winter 2009–2010 (Fig. 6.4f) was described by the herders as a very bad winter. Several snowfalls occurred in October and although there were warm spells, the snow never completely melted, resulting in a snowpack with layers of thawed and refrozen snow.

Adaptive Strategies Based on their experience from the winter of 2007–2008, the herders decided not to use their traditional winter lands. The *siida* Nutti used a plateau, above the tree limit, located in the fall territory. This mountain area had been used in the past by the elder of this winter-group, but it was the first time for the other members. This strategy was successful as the reindeer grazed on the plateau during the entire winter where they had very good grazing conditions. The *siida* Utsi decided to join their herd with those of other *siida* and to use the 'reserve grazing'[2] area located south-east of their traditional grazing area. Here, snow conditions were considered to be better and they managed to get through the winter in this way.

Based on these detailed descriptions of land use by two *siida* over six winters, two conclusions emerged from the co-production workshop. The first is

Table 6.2 *Winter grazing conditions and their consequences as described by the herders during the 2009 workshop*

Winter	Grazing conditions	Adaptive strategies	
		Siida **Utsi**	*Siida* **Nutti**
2004-05	Good	Normal use of the winter grazing area	
2005-06	Good	Normal use of the winter grazing area	
2006-07	Very poor	Silage and pellet feeding of yearlings and use of the fall territory	
2007-08	Very poor	Early gathering of reindeer and grazing in mountain areas	Severe scattering of reindeer during winter
2008-09	Good	Normal use of the winter grazing area	
2009-10	Very poor	Grazing in the 'reserve grazing' area	Grazing in the mountain area

quantitative. Only three winters out of six were considered by the herders to be 'good' or 'normal', while the other three were judged to have offered 'very poor' grazing conditions (Table 6.2). Bad winter or spring seasons have occurred in the past, but they were rare. Today thaw–freeze episodes that prevent reindeer from grazing lichen under the snow are increasingly frequent and are expected to occur even more frequently as climate change advances (SMHI, 2015).

The second conclusion is that land use patterns vary considerably from one winter to the next in response to changing snow and grazing conditions. Variations do not occur in space only but also in the timing of land use. Mapping space-time sequences was thus critical for understanding pasture management. This temporal dynamic is not taken into account in Western mapping, which evaluates pastures through a quantitative account of edible biomass. Herders, on the other hand, think of their pastures more in temporal terms rather than simply as units of space. A given space can feed the herd for a defined time, if snow conditions are excellent, or not feed them at all, if conditions are bad. Between these two extremes, an entire panoply of variants exists, which the herders assess to build an adaptive strategy in space and time. Considering both of these dimensions is essential if we are to understand land use patterns and the constant efforts that herders make to adapt to changing conditions.

Limits of Participatory Mapping for Sámi Management of Uncertainty

Participatory mapping of Indigenous lands to enhance natural resource manage-ment, secure tenure or strengthen cultures is a relatively recent undertaking. Initiated some forty-five years ago, its use has expanded considerably since the

1990s, accompanied by an increased access to technologies such as geographic information system (GIS) (Chapin et al., 2005). Even with participatory methods, however, Indigenous land use mapping often oversimplifies highly complex situations. Rundstrom (1995) voiced concerns about the epistemological gap between Indigenous knowledge systems and GIS, arguing that 'the epistemological system within which GIS is grounded is largely incompatible with the corresponding systems of Indigenous peoples'. While GIS is conceived by Western geographical sciences as an objective methodology that is largely beyond dispute, many authors have criticized its incapacity to accommodate other ontologies and forms of spatiality (Lewis and Woodward, 1998; Ingold, 2000; Dunn, 2007; Roth, 2009). By abstracting and concealing the complex lived spaces of Indigenous peoples, Sletto (2009) speculates that even counter-mapping may reinforce rather than counter state control over Indigenous lands.

Although often used as a management tool, land use mapping perpetuates a perspective that ignores the historical reality of the ancestral Sámi homeland, the dispossession of their territory, and the depth and breadth of Sámi knowledge and resource management capacity. The disparity between the usufruct of reindeer herders and the property rights granted by the State to forest companies is never considered. As a result, an issue of land dispossession is presented as a mere problem of communication that can be solved by adequate mapping of different land uses.

Mapping the 'best available pastures' is a major challenge in a continuously changing subarctic environment. The RHP's purpose was to create a multi-scalar tool, but this tool did not take into account the impermanence and complexity which is at the heart of Sámi ecological knowledge and their management of uncertainty. How can one map 'good' pastures that suddenly become 'bad' when the formation of an ice crust blocks access to the lichen? Or, pastures considered 'poor' that suddenly become the best available choice? As pastures do not remain 'good' indefinitely and under all circumstances, it is impossible for Sámi herders to design a plan for an entire season or yearly cycle and apply it, as foresters do, year after year without change.

The RHP and Its Underlying Paradigms: A Critical Analysis

The initial design of the RHP is based on a set of implicit assumptions that are confronted here with the practices and worldview of reindeer herders.

Carrying Capacity: An Old Illusion

Although the RHP of a community is implemented with reindeer herders, it is underpinned by a classic and long-lived paradigm that continues to be widely held in the animal sciences: carrying capacity. Calculated by assessing the number of grazing animals in relation to the available vegetation, it claims to provide a scientifically determined and optimal number of animals for a given territory, which should be adhered to by the herders. As Ivar Bjørklund (1990: 79) had already understood in 1990 when he wrote about Sámi pasture management and the overgrazing issue in Norway, this paradigm does not take into account the knowledge and management of Sámi herders,

For the biologist this concept is a question of the relationship between animals and pasture. For the pastoralists, however, this concept puts them in the middle of this relation. For them, the carrying capacity of the given pasture is a reflection of their capacity to mediate the relation between herd and pasture. Because of the climatic and biological variations which characterize the yearly cycle of the reindeer, it makes little sense to the herder to define the question of carrying capacity in relation to a certain type of pasture.

Sámi herders know, through their traditional knowledge, that their main task is to adapt themselves and their reindeer continuously to environmental changes. A good pasture may become inaccessible under certain climatic conditions (Roturier and Roué, 2009). It is at a given time and in a given place that herders evaluate the quality of a pasture in relationship with the needs of their herds, through constant monitoring followed by strategic management decisions.

A Static Assessment Based on Western Science

An important difference though, between the RHP project in Sweden and what is done in Norway (Benjaminsen et al., 2015) or elsewhere, is that the RHP does not claim that its purpose is to decrease the number of reindeer to solve an overgrazing problem supposedly caused by the herders. Nevertheless, their attempt to classify once and for all the types of pasture in order to improve management remains very ethnocentric. Based on a linear and one-dimensional relationship between reindeer lichen and pasture quality, without taking into account herders' knowledge and management, it omits the main issue: that of Sámi adaptation to climatic and ecological variability. The effort to become participative, to involve and even empower the herders by training them to use modern technologies, was certainly sincere in the RHP programme. This training in modern techniques was much appreciated, particularly by young herders who need to master additional skills apart from their own. They use them, for instance, to interpret management plans sent to them by forestry companies for information. But these technologies are not based on Indigenous and local knowledge assessment and worldviews. Sámi adaptive monitoring evaluates the evolving status of pastures in relation to environmental conditions during the entire winter. For herders to base their decisions on an essentialist assessment and a tool based on fixed qualities of pastures *per se* would not be a step forward.

A Summer Science

The techniques used in the RHP to assess the quality of pastures in winter combine satellite image analysis and fieldwork by herders, who take GPS-localized pictures while evaluating lichen abundance. Ironically, this information, including the satellite images used to delineate the different areas, needs to be collected in summer when the vegetation cover, particularly ground lichens, is easy to observe. As a result, there is a mismatch between a 'summer science', which determines pasture quality based essentially on botanical criteria, and the 'winter knowledge' of the reindeer herders, based on a continuous monitoring of the dynamic relationship between pastures and the overlying snow cover. Botany and agronomy have not been conceived for the Arctic and Subarctic where snow condition is the determining factor for pasture quality during at least six months of the year. This has not stopped scientists from evaluating land cover on their own and drawing conclusions without any participation by reindeer herders. The failure to recognize the mediating role of reindeer herders between pastures and their herds has resulted in numerous clashes between Western science and Indigenous knowledge. Today, the participatory modality continues to impose conventional Western science, albeit in a less authoritative way, rendering a top-down relationship seemingly more palatable, but no less threatening.

Deconstructing a Concept Borrowed from Protected Areas Management

At the time of the 2009 co-production workshop, the herders had already designated, within the framework of the RHP, 'key areas' on their territories. During the workshop, the herders were encouraged to delineate additional 'key areas' where reindeer had grazed during previous winters. This exercise was particularly instructive with respect to winters with very poor grazing conditions, such as the winter of 2007–2008 (Fig. 6.4d). During difficult winters, the reindeer primarily used grazing areas outside of the key areas that had been previously digitized. The herders explained that, in the winter of 2007–2008, the key areas where ground lichens were most abundant had been 'locked in' by hard layers of snow. As a result, reindeer had to graze in other types of habitats such as bogs and spruce forests, where they could access some ground lichens as well as grasses and arboreal lichens (mainly *Bryoria fuscescens* and *Alectoria sarmentosa*).

There are some little patches of lichen on the ground, where the lichen is high like a coffee cup. But they browse other things in the bogs: tree lichens, crowberries, some grasses ...

There are some tree lichens along this stream. There are some of these little spruces, old ones, everywhere, and they have tree lichens.

L.-E. Nutti during the workshop, 18–19 February 2009

Circumstances such as these particularly challenge the concept of 'key areas', which delimits pastures that cannot be used during winters when snow conditions are 'bad'.

The key areas are valid for a 'normal' winter, in 'normal' pasture conditions. Last year [in winter 2007-08], because of the thaw-freeze, the pasture was locked-in everywhere, and the reindeer were going in all directions. With ice, there was no more pasture, no more key areas ...

L.-E. Nutti during the workshop, 18–19 February 2009

If a key area is valid only under certain circumstances, specifically when conditions are good, how will the RHP help the Sámi to manage, and the foresters to understand Sámi land use management, under real, in other words variable, conditions? Indeed, when digitizing key areas, Sámi herders, trained by scientists, used satellite images upon which lichen-rich habitats were easily identifiable. This methodology resulted in the designation of only the largest lichen-rich pine forests as key areas. However, this approach led the participants astray. Identifying winter pasture in summer, a snow-free season, makes little sense given Sámi knowledge of the importance of winter snow conditions and their variability.

The identified key areas matched well with the areas grazed when grazing conditions were good (Fig. 6.6), 'good' being largely synonymous with what the RHP calls 'normal'. Bad winters, on the other hand, were considered abnormal by the RHP, without anyone being conscious of it. These assumed associations of 'good' with 'normal' and 'bad' with 'abnormal' were all the more tenuous given that bad grazing conditions characterized three of the six winters investigated. In other words, the conditions that were 'abnormal' in the past might be becoming 'normal' in the near future.

When Roué and Roturier pointed out these shortcomings with the concept of key areas as initially defined within the RHP, the herders re-considered their approach to mapping. They designated additional key areas, such as those in the vicinity of corrals, used every year regardless of snow conditions. Indeed, the herders defined a number of additional criteria for the identification of key areas for their herds.

The area where we gather our herd is another type of key area. One could have key areas for pasture quality, key areas for animal categories, key areas depending on snow conditions, key areas depending on their use.

L.-E. Nutti during the workshop, 18–19 February 2009

Figure 6.6 Key areas delineated by the reindeer herders in their RHP and land use of the *siida* Nutti and the *siida* Utsi for the six winters between 2004 and 2010. [A black and white version of this figure will appear in some formats. For the colour version, please refer to the plate section.]

Map by S. Roturier

135

The retrospective study of pasture conditions and strategies during the six previous winters allowed us to conclude that different types of key areas exist for the Sámi. As the workshop demonstrated that normal or average years do not, in fact, exist, herders began to delineate additional key areas that correspond with 'bad' years.

The difference between 'key areas' and 'key areas for bad years' is that in the former there is a continuous lichen-mat, whereas in the latter there are only patches of lichen here and there, and you have tree lichens, and other things too.

L.-E. Nutti during the workshop, 18–19 February 2009

This re-conceptualization led to more in-depth discussions. It was concluded, for example, that key areas could also be specific to particular categories of reindeer such as the yearlings, the weakest members of the herd. The key areas designated for bad winters were, in fact, vital for yearlings even in good winters, as they tended to gather there in good years as well as bad.

What we have just drawn [on the map] is mainly for the years like last year [winter 2007–2008], but it is also valid for the good years as the yearlings, which are the weakest ... part of them go there. When the snow can carry their weight, they are lazy, they do not manage to dig. Usually. the yearlings go to these places, and we leave them there.

P.-O. Kuhmunen during the workshop, 18–19 February 2009

Core Area: A Category That Better Corresponds with Sámi Worldviews

Herders have also expressed their concern that the RHP gives too much importance to key areas. They observed that the category 'core area', in spite of having a lower importance in the hierarchy pre-determined by the RHP methodology, was better adapted to Sámi land use. The core areas, as defined by the RHP (Table 6.1), included all the grazing areas under all of the conditions encountered during the previous winters. More precisely, whereas key areas cover the pastures under good conditions, core areas also include the pastures under bad conditions, including other habitat types that become critically important. This illustrates the shortcomings of a hierarchical approach, which values some categories more than others. Even though this was not the intention of the RHP programme, Sámi herders and other land users became trapped by this simplified approach which misrepresented the complexity of the environment and the needs of their herds. As one herder concluded,

If one looks at the core area, they cover everything that we've talked about. If we would forget the key area [...] But it depends on the value that is given to a core area. With respect to forestry, only the key areas matter, if at all ...

L.-E. Nutti during the workshop, 18–19 February 2009

Herders also expressed their concern that key areas based on so-called normal years were already obsolete given the rapidly changing climatic conditions that they are experiencing today.

From Participatory to Co-produced Reindeer Husbandry Plans

Improvements for the Herders at the Winter-Group Level

The co-production workshop was organized in the midst of the RHP process. As a result, it contributed to improving the RHP through the inclusion of detailed knowledge from herders. Some of the initial shortcomings of the RHP were thus corrected in the final output. At the level of the winter-group, Lars-Evert Nutti modified the boundaries of key areas in their RHP immediately after the co-production workshop. He digitized new key areas in the places where he had observed reindeer grazing during bad years, on the basis of the 2007–2008 winter when reindeer grazed in very different habitats. He also commented that the exercise of recalling and accurately mapping his own patterns of land use reinforced his ability to explain, and even defend, his land use and the strategic importance of certain grazing areas when discussing with forest companies. His partner, Per-Olof Kuhmunen, an elder herder, even proposed to repeat the exercise each year. He argued that, 'If we do it, nobody can contest what we say because what we did is right there in black and white, and why we did it, too.' At the same time, he realized that, 'Of course, it will be completely incomprehensible for somebody who does not know anything about it', emphasizing that a reindeer husbandry plan has no value without herders present to explain it in their own terms.

Improvements to the Reindeer Husbandry Plan

The co-production workshop also stimulated the creation of new RHP modules and training sessions. Per Sandström, who participated in the workshop for the RHP, was responsible for training reindeer herders on making their RHPs until the end of the project in 2014. Following the co-production workshop, he updated his training sessions to include these new orientations. Some ontological gaps in the initial design of the RHP were corrected, such as,

> Conception of the RHP at the winter-group level (*siida* vs the larger reindeer herding community or *sameby*): the category 'General seasonal grazing area' (see Table 6.1), the most inclusive category, is now better defined and identified as the area in which one winter-group maintains its reindeer in winter.

Introduction of the time dimension of grazing and the importance of putting
    clear timestamps on each core and key area: even though this was an option
    under the initial version of the RHP, it was not well understood by herders
    when making their own plans and became clearer in the updated version.

Recognition of inter-annual variability: following the co-production workshop, it
    was recognized that land areas where reindeer graze during 'bad' winters should
    be digitized as key areas, at the same level as the grazing areas for 'good' winters.

Some of the innovations from the co-production workshop were included by RHP
programmers in an updated version of the RHP (Per Sandström, pers. comm.). The
module 'My grazing land history' (*Min Betesmarkshistorik*) was created, directly
inspired from the methods and results of our co-produced research. It enables the
reindeer herders to carry out a detailed mapping of land use during a particular
year, including areas used during different periods within a grazing season. In this
manner, it can also serve as a digital diary for the herders themselves.

A second module named 'My view of other land uses' (*Mina Omvärldsfaktorer*)
has also been inspired by our co-production workshop. It allows reindeer herders
to digitize the impacts of other land uses on reindeer (e.g., 'dangerous ice
conditions close to a dam'), while allowing herders to propose actions to mitigate
negative effects and carry out improvements. Examples of such recommended
measures include 'improving fencing along a railway' or 'thinning a forest stand'.
These new modules are now part of the RHP in all reindeer herding communities.
Lars-Evert Nutti has been a pilot user of these modules and his feedback has
considerably helped improve these tools.

Conclusion

The aim of the RHP was to offer a tool for Sámi reindeer herding communities to
map the most important areas for reindeer grazing, thus facilitating the dialogue
with other land users, in particular, the forestry industry. Mapping Sámi pastoralist
use of the landscape and its chronological, alternative and conditional nature,
however, proved to be a major challenge. In the context of the RHP, the ambition
of the co-production workshop was to convert a two-dimensional representation of
pastures on maps into one with three-dimensions that takes into account snow
cover, and even four dimensions by adding time. To purportedly increase their
capacity to manage the land, or communicate better with other land users, the RHP
should not require Sámi to adopt a simplified representation of their environment
using limited categories, and thus sacrifice their complex multi-dimensional
knowledge. Reindeer husbandry depends on adaptation to a changing climate and
environment, leaving no place for misleading concepts such as 'normal years'.

By deconstructing their science-based GIS categories, the hierarchical classification imposed by the RHP design is shown to be far from appropriate.

The analysis of Sámi pasture management during six consecutive winters revealed a much more diverse and extensive use of the territory than the simplified mapping of key areas, which overemphasize the importance of lichen-rich areas. The herders do not use pure space categories, but rather space-time sequences for the use of different areas under various circumstances. As one herder concluded, 'All these areas are key areas, as the whole area is a key area.'

The maps made by reindeer herders during our workshop cannot wholly escape a simplification process, but they capture a thorough and systematic space-time description of the land used by the entire reindeer herd under various grazing circumstances. They identified links and relationships between grazing areas within the winter territory of a *siida*, a social level of organization that was completely ignored by the RHP programme. Finally, the fact that when completing their RHP, herders can now express their desire for changes by other regional actors (the railway, for instance) is a timid step towards co-management. For now, Sámi herders are not consulted to manage the region in which they have lived for thousands of years, but one can imagine that a better regional integration taking into account the herders' views could emerge in the future.

To conclude, we would like to discuss our title, and our claim to have moved forward from mere participation to a co-production of knowledge. The difference between the initial participative programme and the final co-produced workshop could be seen as a simple progression or a difference in degree. In our view, it is, rather, a difference in nature. Paradoxically, democracy and participative projects give the floor to all 'stakeholders', but do not have the capacity or the will to understand other worldviews, languages or ways of life. Participation applies in all kinds of contexts. For instance, citizen science allows citizens to participate in research projects, and they often do so with much interest. Nevertheless, the deal is clear: as analysis stays in the hands of the scientists, citizens do not have any power over the final result. A planner (as in the RHP) also proposes a format and asks partners to fit in as best they can. The invited partner (here, the Sámi herders) cannot impose their own worldview, as it would prevent them from playing the game according to rules established by the planner. In a more equitable relationship, Sámi herders would simply answer that they can only plan according to circumstances and events, which are never the same each year. That is what herders always first express when an outsider asks them how they plan their week, season or activity. But the desire to play the game compels herders to provide such a simplified image of reality that in the end they trap themselves. When asked to draw a map of their best pastures, or the pastures they use during a 'normal' year, they map the best pastures available during a winter where conditions are ideal. They cannot answer a question that has not been asked, and is even avoided by the planner: do all winters fit this ideal image? This is the normal course of a participatory project. Participation means that the two partners work together, but it does not erase the relations of power. The partner who initiates the project sets the rules. This partner may not be a specialist in the subject being dealt with (here Sámi pastoralism and ethnoecology), but is a specialist in planning and GIS. Participation, in most cases, is full of good intentions but is either naïve or cynical. It works with a lot of hidden assumptions, one being that the persons interviewed are not capable of expressing clearly their own patterns, or worse, are not even capable of planning.

Co-production of knowledge, in our view, has the ambition to equally and respectfully recognize the knowledge of all partners. As the central issue is winter pasture management for reindeer herding, the herders are without any doubt the most knowledgeable. Without this recognition of their expertise, it is impossible to go forward. The specific role of researchers in Indigenous knowledge is to act as mediators between the Sámi and the GIS specialist, with sufficient understanding of both knowledge systems and worldviews to build the bridges needed for co-production. Just as it takes many years for a Sámi to become a reindeer herding expert, researchers in Indigenous knowledge require many years to become acceptable go-betweens: first, through understanding a specialized language, as specific knowledge is conveyed through a terminology and categories that can only be understood in the language and culture in which they are anchored. It also takes time to establish a true dialogue and trust. Beyond interdisciplinarity and even transdisciplinarity, a dialogue is created between partners who have been raised in different cultures and trained in different contexts but who nevertheless agree to pool their efforts, addressing global concerns (e.g., climate change, unsustainable development) in a local context using field-based knowledge from both Indigenous and scientific sources.

Acknowledgements

The co-production workshop benefited from financial support from UNESCO's Local and Indigenous Knowledge Systems (LINKS) programme and from the Laboratory of Eco-anthropology and Ethnobiology of the National Museum of Natural History, Paris. The project continued with support from the BRISK programme, funded by the French National Research Agency (ANR).

Notes

1 The Sámi divide the year into eight seasons, each of our four main seasons being followed by a season considered as a hybrid between the previous and the following one, e.g., winter-spring.
2 The 'reserve grazing' area, located outside of the official borders of the Sirges community, can be used when winter grazing conditions are very poor. To gain access, the community has to apply to the board of the regional County.

References

Batisse, M. 1986. Developing and focusing the biosphere reserve concept. *Nature and Resources*, 22(3): 1–12.

Benjaminsen, T. A., Reinert, H., Sjaastad, E. and Sara, M. N. 2015. Misreading the Arctic landscape: A political ecology of reindeer, carrying capacities, and overstocking in Finnmark, Norway. *Norsk Geografisk Tidsskrift - Norwegian Journal of Geography*, 69(4): 219–229. https://doi.org/10.1080/00291951.2015.1031274

Bjørklund, I. 1990. Sámi Reindeer pastoralism as an Indigenous resource management system in northern Norway: A contribution to the common property debate. *Developmental Change*, 21: 75–86. https://doi.org/10.1111/j.1467-7660.1990.tb00368.x

Chapin, M., Lamb, Z. and Threlkeld, B. 2005. Mapping Indigenous lands. *Annual Review of Anthropology*, 34: 619–638. https://doi.org/10.1146/annurev.anthro.34.081804.120429

Dunn, C. E. 2007. Participatory GIS: A people's GIS? *Progress in Human Geography*, 31: 616–637. https://doi.org/10.1177/0309132507081493

Eira, I. M. G., Jaedicke, C., Magga, O. H., Maynard, N. G., Vikhamar-Schuler, D. and Mathiesen, S. D. 2012. Traditional Sámi snow terminology and physical snow classification: Two ways of knowing. *Cold Regions Science and Technology*, 85: 117–130. https://doi.org/10.1016/j.coldregions.2012.09.004

Ingold, T. 2000. *The Perception of the Environment: Essays on Livelihood, Dwelling & Skill*. London and New York: Routledge.

Jernsletten, N. 1997. Sámi traditional terminology: Professional terms concerning salmon, reindeer and snow. In Gaski, H. (ed.) *Sámi Culture in a New Era: The Norwegian Sámi Experience*. Karasjok: Davvi Girji, pp. 86–108.

Lewis, G. M. and Woodward, D. (eds.) 1998. *The History of Cartography. Vol. 2, Book 3: Cartography in the Traditional African, American, Arctic, Australian, and Pacific Societies*. Chicago, London: University of Chicago Press.

Magga, O. H. 2006. Diversity in Saami terminology for reindeer, snow, and ice. *International Social Science Journal*, 58(187): 25–34. https://doi.org/10.1111/j.1468-2451.2006.00594.x

Riseth, J. Å., Tømmervik, H., Helander-Renvall, E., Labba, N., Johansson, C., Malnes, E., Bjerke, J. W., Jonsson, C., Pohjola, V., Sarri, L. E., Schanche, A. and Callaghan, T. V. 2011. Sámi traditional ecological knowledge as a guide to science: Snow, ice and reindeer pasture facing climate change. *Polar Record*, 47(3): 202–217. https://doi.org/10.1017/S0032247410000434

Roth, R. 2009. The challenges of mapping complex Indigenous spatiality: From abstract space to dwelling space. *Cultural Geographies*, 16: 207–227. https://doi.org/10.1177/1474474008101517

Roturier, S. and Roué, M. 2009. Of forest, snow and lichen: Sámi reindeer herders' knowledge of winter pastures in northern Sweden. *Forest Ecology and Management*, 258: 1960–1967. https://doi.org/10.1016/j.foreco.2009.07.045

Rundstrom, R. A. 1995. GIS, Indigenous peoples, and epistemological diversity. *Cartography and Geographic Information Systems*, 22: 45–57. https://doi.org/10.1559/152304095782540564

Ruong, I. 1964. *Jåhkåkaska sameby. Särtryck ur Svenska Landsmål och Svenskt Folkliv*, 41–158.

Ryd, Y. 2007 [2001]. *Snö: renskötaren Johan Rassa berättar*. [Snow: the reindeer herder Johan Rassa tells]. Stockholm: Natur och kultur.

Sandström, P., Granqvist Pahlén, T., Edenius, L., Tømmervik, H., Hagner, O., Hemberg, L., Olsson, H., Baer, K., Stenlund, T., Göran Brandt, L. and Egberth, M. 2003.

Conflict resolution by participatory management: Remote sensing and GIS as tools for communicating land-use needs for reindeer herding in northern Sweden. *Ambio*, 32: 557–567. https://doi.org/10.1579/0044-7447-32.8.557

Sletto, B. 2009. Special issue: Indigenous cartographies. *Cultural Geographies*, 16: 147–152. https://doi.org/10.1177/1474474008101514

SMHI (Swedish Meteorological and Hydrological Institute). 2015. Meteorological observations. [http://opendata-download-metobs.smhi.se/explore/#]

Swedish Forest Agency. 2001. Skogsbruk och rennäring. [Forestry and reindeer husbandry]. Rapport 8. Skogsstyrelsens förlag.

Swedish Forest Agency. 2003. Projekt Renbruksplan 2000-2002 slutrapport – ett planeringsverktyg för samebyarna. [The Renbruksplan project 2000-2002 final report – a planning tool for reindeer husbandry communities]. Rapport 5. Skogsstyrelsens förlag.

Swedish Forest Agency. 2014. *Renbruksplan – från tanke till verklighet*. [Reindeer Husbandry Plan – from thought to reality]. Rapport 2. Skogsstyrelsens förlag.

Source: http://shop.skogsstyrelsen.se/sv/publikationer/rapporter/renbruksplan-fran-tanke-till-verklighet.html

Figure 2.2 A *bidarki* in its natural habitat during low tide on the lower Kenai Peninsula, Alaska. This individual is about 8 cm long.
Photo by Henry Huntington

Figure 2.3 Interview in Savoonga to document local and traditional knowledge about the Bering Sea ecosystem.
Photo by Henry Huntington

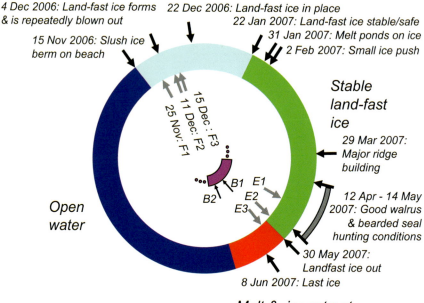

Figure 3.3 Schematic of the seasonal ice cycle at Wales in 2006/07, based on observations by Winton Weyapuk, Jr. (outermost set of black arrows), satellite remote sensing data (inner set of grey arrows), and Alfred Bailey's observations in 1922 (innermost purple circle, thin black arrows correspond to specific observations/photos). Abbreviations: F1 – First ice growing out from shore, F2 – Start of freeze-up offshore, F3 – Ice edge south of Bering Strait; E1 – Ice edge reaches Bering Strait during ice retreat on 7 May 2007, E2 – Ice edge pulls away from Bering Strait to North on 21 May 2007, E3 – Last ice off Wales (8 June 2007); B1 – "Winter conditions still prevailed on June 3, 1922" (Alfred Bailey), B2 – Land-fast ice still in place 16 June 1922 (Alfred Bailey).

Figure 3.4 A slush berm (*qaimġuq*) forming along the shore on 9 November 2007, with slush ice (*qinu*) and small pieces of pancake ice (*nutiġaġuugvik*) moving with the current just off the slush berm.
Photo by Winton Weyapuk, Jr

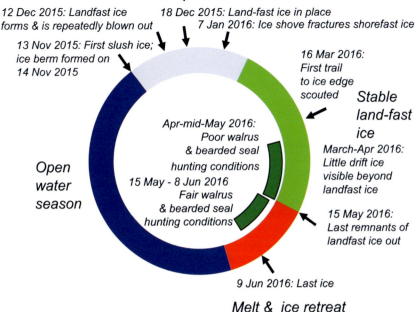

Freeze-up & ice advance

12 Dec 2015: Landfast ice forms & is repeatedly blown out

18 Dec 2015: Land-fast ice in place
7 Jan 2016: Ice shove fractures shorefast ice

13 Nov 2015: First slush ice; ice berm formed on 14 Nov 2015

16 Mar 2016: First trail to ice edge scouted

Stable land-fast ice

March-Apr 2016: Little drift ice visible beyond landfast ice

Apr-mid-May 2016: Poor walrus & bearded seal hunting conditions

15 May - 8 Jun 2016 Fair walrus & bearded seal hunting conditions

Open water season

15 May 2016: Last remnants of landfast ice out

9 Jun 2016: Last ice

Melt & ice retreat

Figure 3.5 Schematic of the seasonal ice cycle at Wales, based on observations by Winton Weyapuk, Jr. in 2015/16.

Figure 3.7 Shorefast ice (*tuaq*) disintegrates, with the remaining broken pieces (*tuwaiġnit*) drifting away. Scattered pack ice (*tamalaaniqtuaq*) has been driven back close to the shore by northern wind on 7 June 2007, with some floes grounded in shallow water (*ikilituat*). The pack ice that has drifted north and then drifted back south by wind-driven current is called *atitiqtuaq*.
Photo by Winton Weyapuk, Jr

(a)

(b)

Figure 4.2 (a) An Utqiaġvik whaling crew moving their hunting equipment out onto the shorefast ice in spring; (b) Utqiaġvik, Alaska, a village of approximately 4,000 people, is shown in springtime with a shelf of shorefast ice along the coast. Photos by Matthew L. Druckenmiller

Figure 4.5 North Slope Borough biologist Craig George (centre) and Utqiaġvik whaler Margaret Opie (right) discussing the location of the main coastal ocean currents off Utqiaġvik during the 2013 workshop.
Photo by Nokimba Acker

Figure 5.5 Page from the 'Wales Sea Ice Dictionary' (Weyapuk and Krupnik, 2012: 54). Caption under photo: Female walrus and a calf (*isavgalik*) are resting on the ice (*nunavait*) in the midst of scattered pack ice (*tamalaaniqtuaq*) interspersed with patches of calm flat water (*quuniq*). The mass of floating pack ice, *sigu*, consists of various types of ice, such as *puktaat* – large floes, *puikaanit* – vertical blocks of ice, *qaŋiłaq* – ice floes with overhanging shelves, *taaglut* – large pieces of darker ice, and *saŋałait* – small floating pieces of dirt ice. May 21, 2007.
Photo and inscribed ice terms by Winton Weyapuk, Jr

Figure 4.6 The many states of Utqiaġvik's coastal waters: (a) a seal hunter on new ice in fall; (b) a deteriorating ice trail in spring with a 'water-sky' (clouds indicating open water) on the horizon; (c) boating in summer fog among drifting ice floes; and (d) a shear boundary between opposing currents during the open water season. Photos (a), (b) and (c) by Matthew L. Druckenmiller; Photo (d) by Hajo Eicken

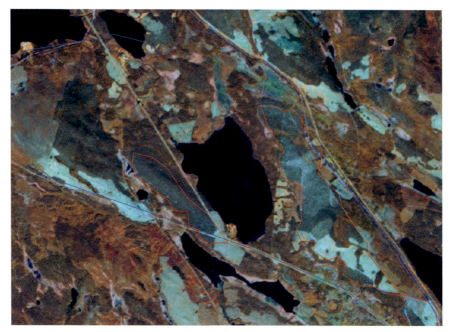

Figure 6.1 The satellite image used as background for digitizing core and key areas for the RHP. The Red/Green/Blue combination display allowed fairly accurate discrimination of forest vegetation. The lichen-rich pine forests are easily identifiable by their blue-grey colour. One core area (blue lines) and two key areas (red lines) are shown on this image.
(Source: Sirges RHP)

Figure 6.5 Artificial feeding of reindeer in late spring.
Photo by S. Roturier

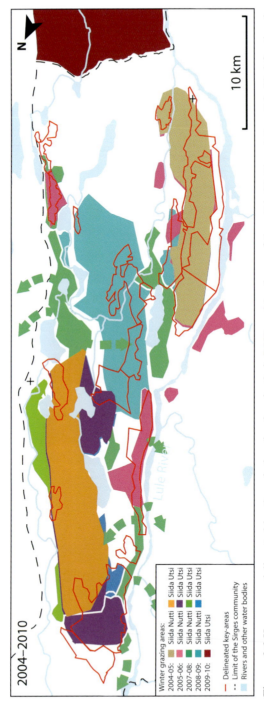

2004–2010

Winter grazing areas:
2004-05: Siida Nutti	Siida Utsi
2005-06: Siida Nutti	Siida Utsi
2007-08: Siida Nutti	Siida Utsi
2008-09: Siida Nutti	Siida Utsi
2009-10: Siida Utsi	

Delineated key-areas
Limit of the Sirges community
Rivers and other water bodies

N

10 km

Lule River

Figure 6.6 Key areas delineated by the reindeer herders in their RHP and land use of the *siida* Nutti and the *siida* Utsi for the six winters between 2004 and 2010.
Map by S. Roturier

Figure 7.2 Reindeer lichen (here mainly *Cladonia stellaris, C. arbuscula and C. rangiferina*) can form extensive mats when they are not outcompeted by ericaceous dwarf shrubs.
Photo by S. Roturier

Figure 7.5 Hans Winsa, Lars-Evert Nutti and Arto Hiltunen prepare the custom-made machine to disperse lichen.
Photo by S. Cogos

Figure 9.5 'Veitau Waqa – the boat lives' - the annual races at Suva Harbour that were initiated by Pacific Blue Foundation as part of efforts to keep traditional cultures alive. By 2014, there were ten smaller *camakau* and two larger *drua* available for the 'Veitau Waqa' event.
Photo courtesy of Aaron March

Figure 9.6 The children challenge each other in the *bakanawa* race during a mid-day break at the annual 'Veitau Waqa-the boat lives' event.
Photo courtesy Aaron March

Figure 11.2 Hanafi Dicko with his herd.
Photo courtesy of H. Dicko

Figure 12.3 Case study of Chizong wine production, Tibetan Autonomous Prefecture, Yunnan, China: (a) French missionaries established a lovely Catholic church in Chizong on the southeastern edge of the Tibetan plateau; (b) the ceiling vault depicts endemic alpine flowers of the eastern Himalaya first described by French botanists; and (c) Tibetan ice wine production won international acclaim.

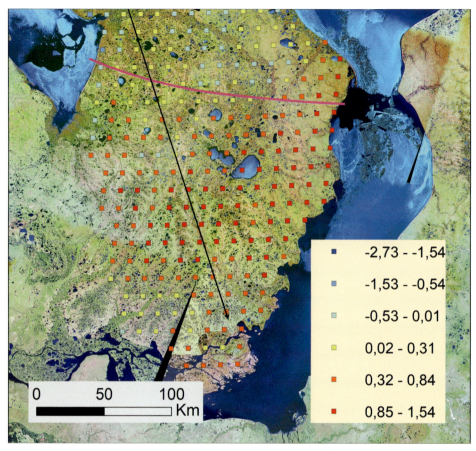

Figure 13.3 ASCAT detected backscatter difference (dB), southern Yamal Peninsula 10 November 2013. The warmer colours (yellow, orange, red) correspond to thicker, more impenetrable ice layers on the snow and/or ground surface. The pink line borders severely iced pasture areas; black arrow indicates reindeer herders' southward migration.

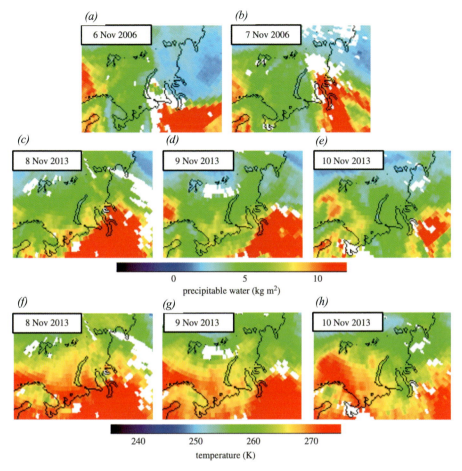

Figure 13.5 AIRS daily total precipitable water (a–b) from 6–7 November 2006 and (c–e) 8–10 November 2013 and 925 hPa temperature from (f–h) 8–10 November 2013 for the Barents and Kara Seas region. White indicates missing data and black outlines the coasts.

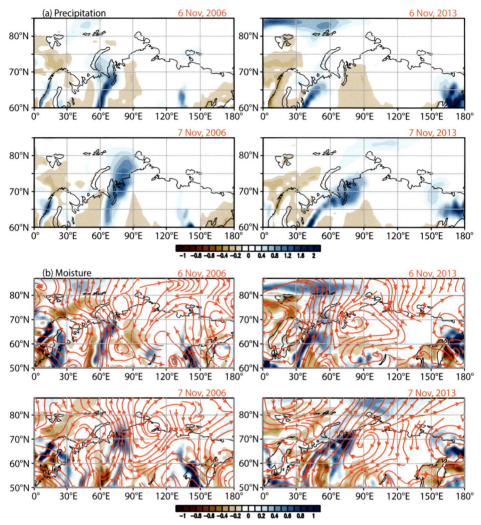

Figure 13.6 (a) Anomalous precipitation (mm) patterns 06–07 November 2006 (left column) and 06–07 November 2013 (right column). Anomalies are determined with respect to monthly averages. (b) Moisture transport (streamline) and convergence of moisture transport (shading: 10-7 s-1) in the 1,000–850 hPa level in 2006 and 2013. Moisture convergence is overall in reasonable agreement with precipitation, suggesting that convergence of increased moisture is primarily responsible for precipitation.

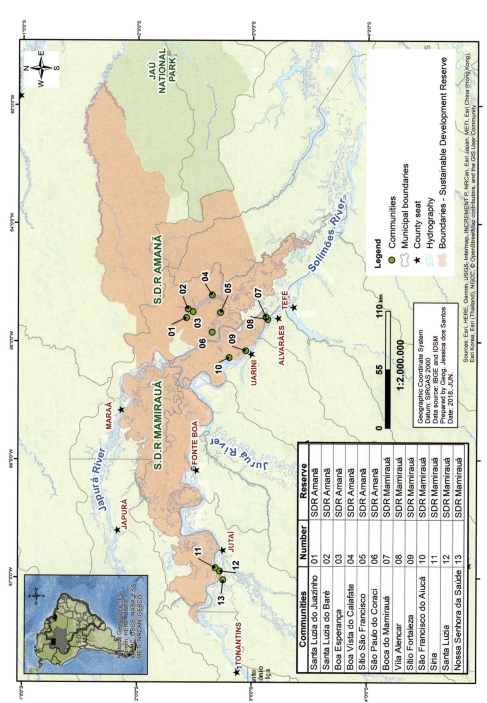

Communities	Number	Reserve
Santa Luzia do Juazinho	01	SDR Amanã
Santa Luzia do Baré	02	SDR Amanã
Boa Esperança	03	SDR Amanã
Boa Vista do Calafate	04	SDR Amanã
Sítio São Francisco	05	SDR Amanã
São Paulo do Coraci	06	SDR Amanã
Boca do Mamirauá	07	SDR Mamirauá
Vila Alencar	08	SDR Mamirauá
Sítio Fortaleza	09	SDR Mamirauá
São Francisco do Aiucá	10	SDR Mamirauá
Síria	11	SDR Mamirauá
Santa Luzia	12	SDR Mamirauá
Nossa Senhora da Saúde	13	SDR Mamirauá

Figure 14.1 Map showing the location of study communities in Mamirauá and Amanã Sustainable Development Reserves, Amazonas, Brazil.

Map credit: Jessica Poliane Gomes dos Santos

Figure 15.1 The Maasai Indigenous peoples of the Kenyan Rift Valley have a rich tradition of oral storytelling as a way of transmitting ILK across generations. Maasai oral histories are vital to the transmission of certain ecological management practices that can be crucial for climate change adaptation and conservation.
Photograph by Joan De La Malla

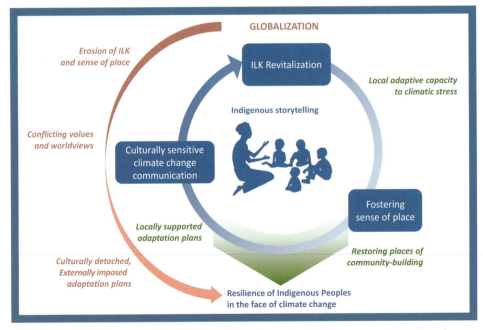

Figure 15.2 Conceptual model describing the role of Indigenous storytelling in climate change adaptation planning. Storytelling is a useful tool for achieving different adaptation goals aimed at fostering social-ecological resilience among Indigenous peoples (blue). Through the promotion of storytelling, such goals can be translated into a series of responses promoting culturally-relevant adaptation (green) while also addressing some of the impacts of globalization upon Indigenous cultures (red).

7

Sámi Herders' Knowledge and Forestry: Ecological Restoration of Reindeer Lichen Pastures in Northern Sweden

SAMUEL ROTURIER, LARS-EVERT NUTTI AND HANS WINSA

Introduction

You can grow pines, you should be able to grow lichen!

This is what a Sámi reindeer herder from the Muonio community in northern Sweden (Fig. 7.1) remembers saying to forest managers from Sveaskog, the Swedish national forestry company, after a final felling had destroyed an important lichen pasture for his reindeer. The incident occurred in 1998 and was the starting point for a long-term research project aimed at restoring reindeer lichen pastures that had been degraded by forestry practices. From the year 2000 to the present day, this research project has evolved to include a Sámi herder, a forest manager and an ethnoecologist. By combining Indigenous knowledge, forest ecology and ethnoecology, a collaborative process of knowledge co-production has emerged. This has encouraged an ambitious agenda aimed at restoring reindeer lichen habitats, including the complex interactions with fire in the boreal forest, and their associated social and ecological components.

This chapter will trace the progress of twenty years of research, examining its different phases. It will also examine the particular factors and specific conditions that were necessary for genuinely co-produced research to finally emerge. By co-production of research we mean, (1) research based on different bodies of knowledge, Indigenous on the one hand and scientific on the other, that complement and reinforce each other; (2) research in which the different partners are equally involved in defining the objectives and can all benefit from the outcomes; and (3) research that cannot be conducted without the complementarity of the knowledge of different partners (Berkes and Jolly, 2001; Pohl et al., 2010; Armitage et al., 2011).

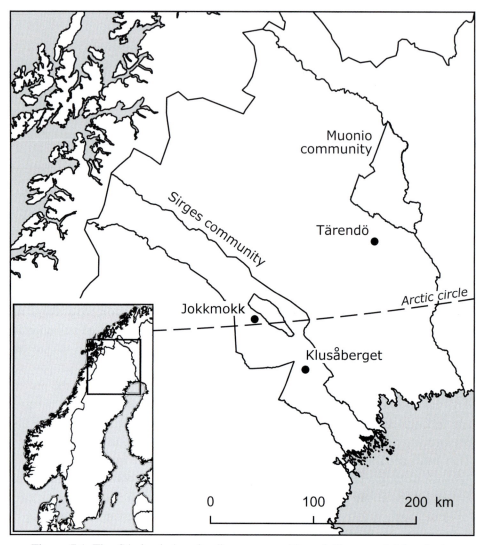

Figure 7.1 The Sámi reindeer herding communities of Sirges and Muonio in northern Sweden.

Land Use Conflict between Forestry and Sámi Reindeer Husbandry: Modalities for Enhancing Dialogue

For centuries, reindeer husbandry has relied on pastures in forestland. Particularly during winter, reindeer (*Rangifer tarandus tarandus*) graze the understorey vegetation, especially ground mat-forming lichen, also known as reindeer lichen, a functional group of terricolous macro lichens mainly belonging to the genus *Cladonia* (Fig. 7.2) that can be very abundant in dry oligotrophic Scots pine (*Pinus*

Figure 7.2 Reindeer lichen (here mainly *Cladonia stellaris, C. arbuscula and C. rangiferina*) can form extensive mats when they are not outcompeted by ericaceous dwarf shrubs. [A black and white version of this figure will appear in some formats. For the colour version, please refer to the plate section.]
Photo by S. Roturier

sylvestris) forests. However, it was recently estimated that, in Sweden, the area of lichen-rich forests has declined by 71 per cent over the last sixty years (Sandström et al., 2016). There may be several intertwined reasons for this decline: infrastructure development such as mines and roads, climate change-induced greening (Cornelissen et al., 2001), locally intensive reindeer grazing (Akujärvi et al., 2014) and finally, commercial forestry (Kivinen et al., 2010). Over the last 150 years, private landowners and forestry companies have exploited the forests for timber and fibre production in numerous ways – from tree regeneration to final harvest – that have heavily and adversely affected reindeer lichen cover. During the second half of the twentieth century, these extractive processes have intensified and the development of forestry in the reindeer husbandry areas of Sweden has become a source of increasing conflict.

From its beginnings in the nineteenth century, commercial forestry has sought to control the main drivers of boreal forest ecosystem dynamics (Esseen et al., 1997). For example, fire, a key element, as we shall see later, in the growth and development of boreal forests was rigorously suppressed in order to protect timber resources. Today's interventionist forest management regimes, which are mainly based on clear-cutting, artificial regeneration of tree stands, shorter rotation age, nitrogen fertilization and planting of exotic species (especially lodgepole pine – *Pinus contorta*), are equally disruptive. Indeed, measures taken to increase forest production by modifying abiotic conditions such as light, temperature, air moisture, soil humidity and fertility can negatively impact reindeer lichen (Kivinen et al., 2010). Exacerbated by other land encroachments, the decline or the degradation of reindeer lichen cover has dramatic consequences for reindeer husbandry. Artificial feeding in the form of food pellets or silage has become more frequent, thus heavily impacting herders' workload and economy.

Interactions between Sámi herders and commercial forest managers were at first largely non-existent, but have gradually intensified and have now become formalized following the development of consultation procedures in 1979. According to the Swedish Forestry Act, 1979, 'consultations' with reindeer herding communities are to be held by forest owners for forestry works that impact an area larger than twenty hectares in the so-called year-round grazing areas, which mainly include summer grazing areas. During consultations, forestry companies, which own and manage large forest stands, must explain to the herding communities the exact areas that will be affected by their intended actions such as clear-cutting, fertilization or forest road construction. Reindeer herders have no formal power (Sandström and Widmark, 2007), but they can engage in negotiations that may postpone forestry works for a couple of years or mitigate the impacts.

These consultation procedures have considerably improved in the last two decades under the impetus of the Forest Stewardship Council Standard, an international environmental certification for wood products adopted by all forest companies in Sweden today, which extended the consultation to winter grazing areas. Even though consultation procedures are not binding on forest companies, they have undoubtedly contributed to an increase in the visibility and importance of communications and discussions between the forestry sector and Sámi herders. In this context, the Swedish Forest Agency (Skogsstyrelsen), the national body responsible for forest administration, led a participatory mapping project from the year 2000 to 2010 which was expected to improve the relationship between reindeer herders and forest managers and facilitate consultations with forestry companies (Sandström et al., 2003; see also Roué et al., Chapter 6).

As an extension of these consultations, another dialogue arena has emerged over the last couple of decades through so-called excursion trips, which are a veritable institution in the forestry sector. They provide forest managers and other forest stakeholders (representatives from the forest administration, scientists, etc.) with the opportunity to reflect on forest management whilst actually in the field. Such excursion trips can also be organized by the forestry sector to meet representatives from reindeer herding communities in order to discuss and exchange views about forest management and its interactions with reindeer husbandry. Although these excursion trips are not a legal requirement, they have become more frequent during recent decades and have undeniably improved the dialogue between parties embedded in different knowledge systems.

This emerging context of dialogue and exchange in the reindeer herding areas has undoubtedly led forest managers to try to resolve the apparent contradiction between sustainability goals such as increased forest production on the one hand, and heightened attention to the interests of reindeer herders on the other. This was formalized, for instance, in 2005 in the Swedish 'National Goals for the Forestry Sector' guidelines that recommend that forest managers pay particular attention to reindeer herding, especially to the important role of lichen-rich pine forests. As a state company, Sveaskog was showcased by the government as a model for the forestry sector. Despite these efforts, the Forestry Act requires, with no ambiguity, that forestry companies and small forest owners meet strict standards of management aimed at maximizing wood production (age of final harvest, forest regeneration, etc.), leaving little room for alternative management that would deviate from this objective (Beland Lindahl et al., 2017).

'Grow Lichen!'

The genesis of the research discussed in this chapter was the provocative demand of the reindeer herder Thomas Sevä, who was devastated to learn about a felling that had destroyed important winter pastures for his reindeer. During an excursion trip in 1998, he challenged Sveaskog, the national forestry company, to grow lichen in the same way that it grows Scots pines. The predicament that Thomas found himself in, however, needs to be described in greater detail: the forestry company and the reindeer herding community had, in fact, agreed upon the felling but the massive destruction of lichen cover was the result of soil scarification carried out at the wrong location. Soil scarification is a standard practice in Swedish forestry to prepare the soil of the clear-cut by mechanically removing the vegetation cover and exposing mineral soil to decrease plant competition and improve pine seed germination or seedling growth. Although he was not in charge of this specific forest area, Hans Winsa, forest manager at Sveaskog, was informed of what had happened and he became involved because the conflict occurred in his home municipality. He accepted Thomas' challenge and, together with Urban Bergsten from the Swedish University of Agricultural Sciences (SLU), raised funds to test the possibility of using natural properties of the lichen to restore areas disturbed by forestry. In 2006, this project became the subject of the doctoral thesis of Samuel Roturier, whose research was mainly funded by Sveaskog.

Whereas most lichens can reproduce either asexually or sexually, reindeer lichen mainly propagates asexually through the dispersal of fragments. The extreme brittleness of the lichens when dry facilitates dispersal, mediated by wind and animals. Therefore, the initial hypothesis tested was that since reindeer lichen reproduces naturally by fragmentation, it should be possible to restore lichen mats after disturbance by directly dispersing or facilitating the dispersal of lichen fragments.

The intriguing results of this innovative research began to emerge after six years of experiments. The research showed that modified, less destructive soil preparation techniques in which fragments of lichen are mixed in the ground layer could facilitate a rapid recovery of the lichen mat within about ten years (Roturier and Bergsten, 2006; Roturier et al., 2011). By comparison, when conventional, more destructive techniques are used, only a layer of mineral soil remains and it can take between 50 and 100 years for the lichen mat to recover. The results demonstrated that fragments of lichen scattered over disturbed forest soil could establish and grow (Roturier et al., 2007; Roturier and Bergsten, 2009). In summary, the possibility of artificially dispersing lichen fragments that would build a new mat or accelerate the recovery of a degraded one was demonstrated. Thus, the restoration of lichen grounds could be achieved with a modest investment of time and money. This outcome was appealing to forest managers because it had the potential to significantly reduce forestry impacts on reindeer herding without changing standard practices, such as mechanical soil preparation. For reindeer herders, this result was positive as it could partially mitigate a major impact on grazing grounds. For the researchers involved in the project, it was important that the capacity to restore would not provide a pretext for further degradation, a classic dilemma debated in ecological restoration (e.g., Young, 2000). In addition, it was important to demonstrate that this work could bring tangible benefits to the herders and did not simply remain an interesting experiment without further application.

Towards Co-produced Research: Coupling Lichen Dispersal with Prescribed Burning

In June 2008, an excursion trip for staff of the Swedish Forest Agency, the main Sámi reindeer-herding organization, Sámiid riikkasearvi, and forest researchers was organized by Hans Winsa and Samuel Roturier to present the initial results of the lichen restoration programme. The excursion included a visit to a burnt clear-cut area where Hans Winsa and Samuel Roturier had carried out the first test of lichen spreading two years earlier. This test had not been successful, probably owing to a combination of factors, including problems with the conditions of lichen storage and a dispersal too soon after the fire. Lars-Evert Nutti, a Sámi herder who participated in the trip, suggested trying again in an area close to a hill named Klusåberget that had burnt in 2006. This fire, known as the Bodträskfors fire, was one of the largest in modern Swedish history, burning about 18 km^2 of the winter grazing territory of Lars-Evert's community. This proposal, welcomed with enthusiasm, led to the Klusåberget experiment.

As a forest manager, Hans Winsa was convinced of the high potential of prescribed burning for forest regeneration and was equipped with the knowledge and know-how associated with this practice. In the 1950s, prescribed burning was the dominant silvicultural measure used to prepare stands for forest regeneration. The practice died out in the 1970s, however, owing to the costs and risks of prescribed burning and the expanding use of forestry machinery (see Cogos et al., 2020 for a more detailed history of prescribed burning in Swedish forestry). Hans Winsa's home district, Tärendö, remains the only one in Sweden where this burning tradition has continued without interruption.

In northern Sweden, the incidence of fire has decreased from the end of the nineteenth century as a result of laws to protect timber resources that prohibit anthropogenic fire, as well as the increased efficiency of wildfire suppression techniques (Granström and Niklasson, 2008). Historically, the fire return interval (i.e., the time between two fires) in Swedish dry pine forests ranged between 20 and 100 years (Zackrisson, 1977; Engelmark, 1984), a much higher frequency than today's expected return intervals of hundreds of years (Niklasson and Granström, 2000; Wallenius, 2011). The prolonged exclusion of fire alters below and above ground properties that induce, especially on mesic sites, an understorey vegetation of feather mosses and ericaceous dwarf shrubs that outcompete ground lichens (Nilsson and Wardle, 2005). As a result, while fire burns out reindeer lichen for several decades, fire suppression contributes in the long-term to a decrease in lichen-rich forests at landscape scale.

Sámi reindeer herders are aware of the influence of past fires on boreal forests. Based on their observations of burnt stumps or trees in the old forest, they believe that each forest has burnt at least once. Also, the Sámi word *roavvi* refers to a heathland that herders usually consider as good pasture, but in some regions it translates as 'a place that was once burnt' (Nielsen and Nesheim, 1979; see also Cogos et al., 2019 for a detailed analysis of *roavvi*). Yet, unlike in North America (Lewis and Ferguson, 1988; Anderson, 2005) or Australia (Langton, 1998), where Indigenous peoples' use of fire in the landscape has been extensively documented, there is no known record of fire management being practised by the Sámi people (but see Hörnberg et al., 2018). However, increasing fire activity was clearly established during the eighteenth and nineteenth centuries under the influence of settlers (Niklasson and Granström, 2000).

In the long run, the re-introduction of fire as a disturbance agent in Fennoscandian forests through controlled fire (including prescribed burning prior to forest regeneration and conservation fire for biodiversity objectives) or wildfire ensures the creation of habitat suitable for ground lichen in the landscape (Halme et al., 2013). That said, the final successional phase dominated by ground lichens is reached ca. 100 years after burning, with mat-forming reindeer lichen species only dominant in the ultimate stages (Morneau and Payette, 1989; Ahti and Oksanen, 1990). Combining burning with artificial dispersal of reindeer lichen might accelerate the return of reindeer lichen and maintain areas of lichen-rich pastures, while restoring a key process of boreal forest functioning.

In Klusåberget, Sveaskog provided logistical assistance and, together with Samuel Roturier, found the necessary financial support to undertake this experiment. Lars-Evert Nutti offered the use of his home to prepare the lichen material. During the summer of 2008, the experimental design was discussed and dispersal areas in the field were selected. It was decided that lichen would be dispersed at three different sites following a gradient of fire severity, which could be determined from the amount of organic layer left after the blaze. This was an important feature as it offered different conditions for lichen re-establishment. Also included in the design was the dispersal of lichen during two different seasons, in late summer and in late winter, to test the practicalities of dispersing on snow. Fragmented lichen was dispersed over half a hectare in September 2008, and over another half hectare in March 2009. In total, about 2,500 kg of lichen were dispersed. Lars-Evert Nutti suggested that young herders from his *siida*[1] should be involved in the preparation and fragmentation of lichen, as he wanted them to be trained in this process. The manager responsible for the Sveaskog district participated in the fieldwork in order to keep track of the experiment (Fig. 7.3).

Following dispersal, the establishment and growth of lichen fragments was surveyed and inventoried on three occasions. Seven years later, the results were beyond expectations. There was a continuous increase in lichen over time, and after seven years, lichen was established, on average, on 75 per cent of the experimental plots treated. During the same time period, no lichen had colonized the control, that is, the burnt surface alongside the experimental dispersal area. At the site where standing trees were retained, lichen formed a well-established mat with a significantly higher lichen occupancy and abundance than in open, more severely burnt sites. Lichen that was dispersed in late summer exhibited higher

Figure 7.3 Lars-Evert Nutti along with Per-Erik Nordqvist from Sveaskog disperse lichen with spades at Klusåberget in March 2009.
Photo: S. Roturier

abundance and occupancy than lichen dispersed in late winter (see Roturier et al., 2017 for the complete results).

In September 2015, representatives from the forestry sector and from all Sámi communities of Sweden were invited to participate in an excursion to the Klusåberget experimental site (Fig. 7.4). More than fifty people participated in this trip. Foresters and herders, sometimes from the same district, had the opportunity to discuss and comment on the experiment with respect to their own region. Since then, information about the work has spread and several forest managers and reindeer herders have asked for advice on how to reproduce the experiment on their land.

The design and execution of the field experiment at Klusåberget was a pioneering outcome that was co-produced by partners from very different backgrounds in an equitable relationship. Perhaps the most unexpected, yet significant, result of the experiment was that it triggered a wider, long-term process that transformed all of our roles, creating an even more balanced interaction. And to some degree our respective communities were changed as well.

Figure 7.4 Excursion to Klusåberget with Sámi herders and staff from forestry companies in September 2015.
Photo by S. Cogos

Sharing a Common Understanding for a Common Goal

Following the completion of Samuel Roturier's Ph.D. on the linkages of Sámi herders' knowledge and forestry (Roturier, 2009), Hans Winsa was eager to pursue further research that could improve forest management and its relationship with reindeer husbandry. During the winter of 2010, he acquired funding from Sveaskog to establish new field experiments and launched a process to involve members of Sámi herding communities more deeply in the work, beginning with their identification of the issues to be addressed. Not surprisingly, this worked well in Lars-Evert Nutti's community of Sirges, as he was centrally involved. A series of meetings and informal discussions took place during the following years, at the initiative of various partners. Very quickly, the key issue that emerged was how to apply the accumulated results and knowledge to operational forestry. To this end, forest stands in Lars-Evert Nutti's traditional grazing area were identified to test and further experiment with innovative lichen management. In total, four experiments were jointly designed and implemented between 2008 and 2010 in the herding community of Sirges with the objective of collecting evidence for management solutions that could be applied by forest managers and reindeer herders alike. This process led to an unusual circumstance: an equitable partnership in the midst of a wider landscape of land use conflict where, on the whole, relations remained deeply skewed. Mutually agreeing to research objectives and experimenting in the field to acquire new, shared knowledge, and then reflecting on concrete applications with a transdisciplinary approach, created a climate of mutual trust and respect that nurtured the work.

This process also reversed conventional epistemological realms and roles. Lars-Evert Nutti took lead responsibility for numerous experiments; he contributed to the dissemination of information about the project, especially at Klusåberget, in his own community and in other communities in Jokkmokk in northern Sweden. He also reached out well beyond Jokkmokk; for instance, to the Central Consultation Group, a body that engages reindeer herders and forestry representatives from the entire reindeer herding area. Through this process of empowerment, Lars-Evert has taken the lead in showcasing results from the Klusåberget field experiment on reindeer lichen management, in the same manner as forestry companies, during their field excursions, showcase the effects of silvicultural operations on tree growth. On several occasions, he invited Samuel Roturier to participate in meetings and asked for help to reinforce ecological premises that he raised in his presentations. During this process, he also improved his expertise in the field of forestry. When Hans Winsa shared his experience on estimating the natural regeneration of pine saplings in the field, Lars-Evert seized this opportunity to argue against soil scarification on purely silvicultural grounds: if natural regeneration is sufficient then scarification becomes unnecessary and uneconomical.

Similarly, the knowledge co-production process induced changes within the Swedish forestry company, especially in the district of Tärendö, the home territory of Hans Winsa. Since 2008, Hans Winsa and Arto Hiltunen, a forest manager from Tärendö, have engaged in testing reindeer lichen dispersal, often at their own initiative. Technical discussions about how to disperse lichen have become a familiar topic around the coffee table in the office. Forest managers have even begun to work on developing a machine that fragments and disperses bags of lichen on a large scale, supervising an engineering student on this topic (Fig. 7.5; Krekula, 2007).

Yet, the most interesting change in the future might be the diffusion of the idea of using prescribed burning in the best interest of reindeer lichen regeneration, along with the associated knowledge within the company. Indeed, the district of Tärendö may well take the lead in this respect. Although it is an exception to standard practice in Swedish forestry, prescribed burning for site preparation to improve forest regeneration after clear-cutting is widely used in Tärendö. This technique offers an alternative to more destructive methods such as mechanical soil preparation. Annually, local foresters burn as much as 4 km^2 or about half of the total area burnt by the company throughout Sweden. Such an operation is demanding and requires high levels of technical expertise, from the planning of felling to the control of smouldering debris. Although totally non-existent in other

Figure 7.5 Hans Winsa, Lars-Evert Nutti and Arto Hiltunen prepare the custom-made machine to disperse lichen. [A black and white version of this figure will appear in some formats. For the colour version, please refer to the plate section.]
Photo by S. Cogos

places today, the Sveaskog staff members in Tärendö have maintained a 'fire culture' and have consciously nurtured it among the crew for many years. The local landscape is also well-suited to this practice, with many wet areas that serve as natural fire breaks. Encouraged by their good results in tree regeneration, the Tärendö district continues to apply prescribed burning whenever possible.

Finally in 2011, Hans Winsa, Lars-Evert Nutti and Samuel Roturier asked the fire crew of Tärendö to plan and conduct a prescribed burning in the district of Jokkmokk for one of the model experiments. The objective was two-fold: to benefit from their skills, and to train foresters from another district who are not familiar with controlled burning. The added dimension of managing a resource other than trees, namely reindeer lichen, allowed forest managers to expand their interests beyond the usual realm while still mobilizing their restoration knowledge and expertise.

Lessons for Knowledge Co-production

The co-production of knowledge on ecological restoration of reindeer lichen pastures is dependent on the dynamic and complex endeavours of a number of different actors. It involves the birth of a community of practice (Wenger, 1998). Initiated by forest managers and forest ecologists upon the suggestion of a Sámi herder, it generated new questions that required an environmental anthropologist to be integrated into the project team. It became transdisciplinary when Sámi herders joined the ranks, thus combining science and Indigenous knowledge to address a problem which neither could solve on its own. The complex challenge of restoring reindeer lichen pastures in lands under forest production required knowledge from all of these sources to address the socio-ecological trade-offs between timber production and reindeer grazing. It fostered a collaborative learning process through experimentation using controlled field experiments that were jointly designed, established and surveyed. This provided an agreed upon and shared framework directed towards a shared goal, an essential requirement for the co-production of new knowledge. Finally, it set the stage for transferring this new co-produced knowledge between and within different Indigenous and professional communities, thus enlarging the initial community of practice.

The co-produced research was created through the establishment of field experiments that were rigorously designed to be analysed statistically. The formal 'products' of this research thus did not differ from other scientific experiments. This is an advantage in academic science, as field experiments are a common way of acquiring new knowledge. As both forest managers and reindeer herders became familiar with this approach, the research could follow methodologies and norms of validation that were well established and agreed upon by all of the partners. Objectives, motivations and issues to be addressed were deliberated upon among the partners, sometimes during formal meetings but more often during field trips. Co-produced research using an experimental approach had two other advantages: it sought new knowledge and at the same time tested innovative forms of management.

A major consequence of the experimental approach was its effect on rebalancing the power relations between the two partners: forest managers and Sámi herders. As landowners, forest companies have the right and, indeed, the duty to manage their forests according to forestry law, whereas Sámi only have land use rights. Therefore, a clear imbalance exists. Unbalanced relations can also exist between scientists and reindeer herders: some Sámi do not hesitate to recall occasions when they helped scientists with research projects that had no local returns but only served to further academic careers. In this co-produced research, all knowledge, experience and economic interests of the various partners were at the same level.

Most importantly, this co-produced knowledge brought clear benefits to all parties; being involved in the research process from the beginning ensured that the results were directly available to both forest managers and Sámi herders. Indeed, field-based forest experiments were established for the express benefit of Sámi reindeer husbandry. For Sámi herders, the knowledge gained can be used in general discussions with representatives from forestry, or even in negotiations during future consultation procedures. Using co-produced knowledge, as generated here, the Sámi could assert their management role on an equal footing with others, making specific proposals to control and shape their environment. The inclusion of Lars-Evert Nutti as a co-author in the scientific publication that resulted from the Klusåberget experiment (Roturier et al., 2017) was important in this respect. We believe this to be an essential point worth emphasizing: Sámi reindeer herders are deeply knowledgeable about how to manage the boreal forest in ways that match their pastoral objectives, that is, they have a profound understanding of the processes that influence the extent and the functionality of pastures. While they put this knowledge into practice in their day-to-day planning and activities, their management role continues to be largely disregarded by the State and other landowners. Our work was thus important in extending benefits to a previously marginalized group, but the results were also produced according to exacting scientific standards, meaning that it had enough legitimacy to convince traditional forest managers of its utility: people who can then transmit the knowledge we created collectively through their own conventional channels.

Being on an equal level is certainly a prerequisite for knowledge co-production but it does not necessarily imply that dialogue between the different knowledge systems is straightforward. On the contrary, bridges must be built between the knowledge systems. A common pitfall one can hear expressed by both Sámi herders and forest managers is, 'We don't talk the same language.' This is literally true since many Sámi herders speak Samigiela, and for older herders, Swedish is often only their second language. But what is also meant here is that certain words do not carry the same meaning for different stakeholders, even when the Sámi do speak Swedish. We have illustrated it by considering the complex and nuanced meaning of the Sámi word *guohtun*, which is usually translated as *bete* (pasture) in Swedish, but refers in actual fact not only to the presence of fodder for reindeer but also its accessibility underneath the snowpack in winter (Roturier and Roué, 2009).

Even more important, however, is the fact that Sámi herders and foresters do not share the same worldviews and epistemologies. Probably none of this would have been possible without a finely detailed understanding of the differences between the semantic categories and worldviews of the different partners. It was critical for forest ecologists and scientists to understand Indigenous representations of pastures in order for them to accept Sámi reindeer herders as key partners in tackling the complexities inherent to the challenge of restoring reindeer pastures. In the Klusåberget experiment, reindeer husbandry provided the long-term goals of the restoration in accordance with the complex dynamics of reindeer grazing and pastures; forestry provided expertise about forest regeneration and fire technology; and restoration ecology provided the framework for experimenting with innovative restoration strategies.

Initially, the intermediary position of Samuel Roturier was critical in many respects such as building trust and establishing interdisciplinary contacts; this allowed for the understanding of the distinct knowledge and worldviews of both communities. Arriving as a foreign student and an outsider in this century-long Swedish management conflict, Samuel Roturier played the part of inoffensive learner that matched rather well with the requirements of this function. In addition, Lars-Evert Nutti and Hans Winsa had already built bridges between their respective worlds. Lars-Evert Nutti had participated each year in many consultations, and in addition he was sometimes hired in the summer for forestry works,[2] giving him the opportunity to reflect on the multiple effects of forestry on reindeer pastures.

Finally, as the objectives of the co-produced research were of great importance in the process since it aimed at restoring the land to better health, the potential benefits of a stimulating and tangible project were immediately apparent to all. At the beginning of the ethnographic fieldwork, when Lars-Evert Nutti and Samuel Roturier met for the first time for an interview, Lars-Evert Nutti immediately showed great interest in the experimental work in restoring lichen pastures and in the possible beneficial consequences it could have for Sámi reindeer husbandry. The interview quickly became much more than a researcher asking questions and a herder responding: rather, it was a productive dialogue.

Concluding Remarks

This project has contributed to identifying and filling a significant knowledge gap that will eventually help to promote the wider restoration of reindeer lichen pastures in productive forestlands. One step further would be to initiate a form of adaptive co-management, which permits stakeholders to share the management responsibilities on a more permanent basis and to learn from their respective actions. The emergence in recent years of the concept of *renskötsel anpassade skogsbruk* (forestry adapted to reindeer husbandry) suggests that Swedish forestry may wish to adjust its priorities to adapt to reindeer husbandry. If this is the case, then it should acknowledge that Sámi reindeer herders have long incorporated elements of forestry into their knowledge system and that they can act as experts in managing forest stands in ways that benefit all.

At present, forest managers face a gap that is both regulatory and scientific. The Swedish Forestry Act claims that forest management should take care of reindeer husbandry interests, yet at the same time it does not allow the management of forests to deviate from a pre-defined trajectory that aims at maximizing wood production. This is not without consequences to knowledge production. Although the effects of forestry on reindeer husbandry are now perfectly well established, there is a considerable lack of research about alternative forms of management that could fulfil the objectives of both timber and pasture production (Beland Lindahl et al., 2017). It is interesting that testing new forms of management involves the kind of compromises that are sometimes made by reindeer herders and forest managers during consultation procedures. Although *ad hoc* field experiments of the kind discussed in this chapter are rare, often at the initiative of engaged individuals, and slow-moving because they do not match the requirements of the Forestry Act, the framework for partnership is already in place.

Co-production of knowledge and adaptive co-management will also be important in the future with respect to fire, and there is no doubt that policies towards fire regimes will evolve during the twenty-first century (Eckerberg and Buizer, 2017). For well over 100 years, fire has been absent from boreal ecosystems, and this has had negative consequences, not only for reindeer husbandry but also for biodiversity and ecosystem functioning (Gustafsson et al., 2019). Since the 1990s, however, the issue of fire has been debated in the forestry sector through the lens of conserving threatened fire-dependent species, and today these debates also consider how fire affects reindeer husbandry (Cogos, 2020). Additionally, boreal forest ecosystems are already facing the impacts of climate change which will further affect fire regimes with consequences that are difficult to predict, as shown by the extreme fire season during the summer of 2018. However, in order to adapt to this unprecedented situation, new parameters for the reintroduction and management of fire regimes should be defined in collaboration with all the relevant partners, including forest managers, biodiversity conservationists and Sámi reindeer herders.

Acknowledgements

This research was funded by several donors, mainly Sveaskog, Sweden; the Programme for Interdisciplinary Research in Ecological Engineering of the CNRS (French National Centre for Scientific Research), France; and the French National Research Agency (ANR)-funded programme BRISK (Bridging Indigenous and Scientific Knowledge in the Arctic) (ANR-12_SENV-0005).

Notes

1 The *siida* is the smallest social unit in Sámi society.
2 These works included, especially, pre-commercial thinning, a silvicultural operation that enables the selection of individual trees to be left for production. In addition, it significantly clears the stand, improving the growing conditions for reindeer lichen.

References

Ahti, T. and Oksanen, J. 1990. Epigeic lichen communities of taiga and tundra regions. *Vegetation*, 86: 39–70. https://doi.org/10.1007/BF00045134

Akujärvi, A., Hallikainen, V., Hyppönen, M., Mattila, E., Mikkola, K. and Rautio, P. 2014. Effects of reindeer grazing and forestry on ground lichens in Finnish Lapland. *Silva Fennica*, 48: 1–18. https://doi.org/10.14214/sf.1153

Anderson, M. K. 2005. *Tending the Wild: Native American Knowledge and the Management of California's Natural Resources*. Berkeley: University of California Press.

Armitage, D. R., Berkes, F., Dale, A., Kocho-Schellenberg, E. and Patton, E. 2011. Co-management and the co-production of knowledge: Learning to adapt in Canada's Arctic. *Global Environmental Change*, 21: 995–1004. https://doi.org/10.1016/j.gloenvcha.2011.04.006

Beland Lindahl, K., Sténs, A., Sandström, C., Johansson, J., Lidskog, R., Ranius, T. and Roberge, J. M. 2017. The Swedish forestry model: More of everything? *Forest Policy and Economics*, 77: 44–55. https://doi.org/10.1016/j.forpol.2015.10.012

Berkes, F. and Jolly, D. 2001. Adapting to climate change: Social-ecological resilience in a Canadian Western Arctic community. *Conservation Ecology*, 5(2): 18.

Cogos, S. 2020. *Fire, People and Reindeer in the Boreal Forest. The Role of Fire in the Historical and Contemporary Interactions between Sami Reindeer Herding and Forest Management in Northern Sweden*. Doctoral thesis, Swedish University of Agricultural Sciences & Paris Saclay University.

Cogos, S., Östlund, L. and Roturier, S. 2019. Forest fire and Indigenous Sami land use: Place names, fire dynamics and ecosystem change in northern Scandinavia. *Human Ecology*, 47: 51–64. https://doi.org/10.1007/s10745–019-0056-9

Cogos, S., Roturier S. and Östlund, L. 2020. The origins of prescribed burning in Scandinavian forestry: The seminal role of Joel Wretlind in the management of fire-dependent forests. *European Journal of Forest Research*, 139: 393–406. https://doi.org/10.1007/s10342–019-01247-6

Cornelissen, J. H. C., Callaghan, T. V, Alatalo, J. M., Michelsen, A., Graglia, E., Hartley, A. E., Hik, D. S., Hobbie, S. E., Press, M. C., Robinson, C. H., Henry, G. H. R., Shaver, G. R., Phoenix, G. K., Gwynn Jones, D., Jonasson, S., Chapin, F. S., Molau, U., Neill, C., Lee, J. A., Melillo, J. M., Sveinbjörnsson, B. and Aerts, R. 2001. Global change and arctic ecosystems: Is lichen decline a function of increases in vascular plant biomass? *Journal of Ecology*, 89: 984–994. https://doi.org/10.1111/j.1365-2745.2001.00625.x

Eckerberg, K. and Buizer, M. 2017. Promises and dilemmas in forest fire management decision-making: Exploring conditions for community engagement in Australia and Sweden. *Forest Policy and Economics*, 80: 133–140. https://doi.org/10.1016/j.forpol.2017.03.020

Engelmark, O. 1984. Forest fires in the Muddus National Park (northern Sweden) during the past 600 years. *Canadian Journal of Botany*, 62: 893–898. https://doi.org/10.1139/b84–127

Esseen, P. A., Ehnström, B., Ericson, L. and Sjöberg, K. 1997. Boreal forests. *Ecological Bulletin*, 46: 16–47.

Granström, A. and Niklasson, M. 2008. Potentials and limitations for human control over historic fire regimes in boreal forest. *Philosophical Transactions of the Royal Society of London*, 363: 2353–2358. https://doi.org/10.1098/rstb.2007.2205

Gustafsson, L., Berglind, M., Granström, A., Grelle, A., Isacsson, G., Kjellander, P., Larsson, S., Lindh, M., Pettersson, L. B., Strengbom, J., Stridh, B., Sävström, T.,

Thor, G., Wikars, L. O. and Mikusiński, G. 2019. Rapid ecological response and intensified knowledge accumulation following a north European mega-fire. *Scandinavian Journal of Forest Research*, 34: 234–253. https://doi.org/10.1080/02827581.2019.1603323

Halme, P., Allen, K., Aunins, A., Bradshaw, R. H. W., Brūmelis, G., Čada, V., Clear, J. L., Eriksson, A.-M., Hannon, G., Hyvärinen, E., Ikauniece, S., Iršėnaitė, R., Jonsson, B. G., Junninen, K., Kareksela, S., Komonen, A., Kotiaho, J. S., Kouki, J., Kuuluvainen, T., Mazziotta, A., Mönkkönen, M., Nyholm, K., Oldén, A., Shorohova, E., Strange, N., Toivanen, T., Vanha-Majamaa, I., Wallenius, T., Ylisirniö, A.-L. and Zin, E. 2013. Challenges of ecological restoration: Lessons from forests in northern Europe. *Forest Ecology and Management*, 167: 248–256. https://doi.org/10.1016/j.biocon.2013.08.029

Hörnberg, G., Josefsson, T., DeLuca, T. H., Higuera, P. E., Liedgren, L., Östlund, L. and Bergman, I. 2018. Anthropogenic use of fire led to degraded scots pine-lichen forest in northern Sweden. *Anthropocene*, 24: 14–29. https://doi.org/10.1016/j.ancene.2018.10.002

Kivinen, S., Moen, J., Berg, A. and Eriksson, Å. 2010. Effects of modern forest management on winter grazing resources for reindeer in Sweden. *Ambio*, 39: 269–278. https://doi.org/10.1007/s13280–010-0044-1

Krekula, K. J. 2007. *Technical Possibilities for Artificial Dispersal of Reindeer Lichen.* Master's thesis. Department of Forest Resource Management, 165. SLU.

Langton, M. 1998. *Burning Questions: Emerging Environmental Issues for Indigenous Peoples in Northern Australia. Centre for Indigenous Natural and Cultural Resource Management.* Northern Territory University, Darwin.

Lewis, H. T. and Ferguson, T. A. 1988. Yards, corridors, and mosaics: How to burn a boreal forest? *Human Ecology*, 16: 57–77. https://doi.org/10.1007/BF01262026

Morneau, C. and Payette, S. 1989. Postfire lichen-spruce woodland recovery at the limit of the boreal forest in northern Quebec. *Canadian Journal of Botany*, 67(9): 2270–2782. https://doi.org/10.1139/b89–357

Nielsen, K. and, Nesheim, A. 1979. [1932–1962] *Lappisk (Samisk) Ordbok – Lapp Dictionary*. Vol. I-V. Universitetsforlaget, Oslo: Instituttet for sammenlignende kulturforskning.

Niklasson, M. and Granström, A. 2000. Numbers and sizes of fires: Long-term spatially explicit fire history in a Swedish boreal landscape. *Ecology*, 81: 1484–1499. https://doi.org/10.1890/0012-9658(2000)081[1484:NASOFL]2.0.CO;2

Nilsson, M. C. and Wardle, D. A. 2005. Understorey vegetation as a forest ecosystem driver: Evidence from the northern Swedish boreal forest. *Frontiers in Ecology and the Environment*, 3: 421–428. https://doi.org/10.1890/1540-9295(2005)003[0421:UVAAFE]2.0.CO;2

Pohl, C., Rist, S., Zimmermann, A., Fry, P., Gurung, G. S., Schneider, F., Speranza, C. I., Kiteme, B., Boillat, S., Serrano, E., Hadorn, G. H. and Wiesmann, U. 2010. Researcher's roles in knowledge co-production: Experience from sustainability research in Kenya, Switzerland, Bolivia and Nepal. *Science and Public Policy*, 37(4): 267–281. https://doi.org/10.3152/030234210X496628

Roturier, S. 2009. *Managing Reindeer Lichen during Forest Regeneration Procedures: Linking Sami Herders' Knowledge and Forestry*. Doctoral thesis. Swedish University of Agricultural Sciences, 84.

Roturier, S. and Bergsten, U. 2006. Influence of soil scarification on reindeer foraging and damage to planted Scots pine (*Pinus sylvestris*) seedlings. *Scandinavian Journal of Forest Research*, 21(3): 209–220. https://doi.org/10.1080/02827580600759441

Roturier, S. and Bergsten, U. 2009. Establishment of *Cladonia stellaris* after artificial dispersal in an unfenced forest in northern Sweden. *Rangifer*, 29(1): 39–49. https://doi.org/10.7557/2.29.1.208

Roturier, S. and Roué, M. 2009. Of forest, snow and lichen: Sami reindeer herders' knowledge of winter pastures in northern Sweden. *Forest Ecology and Management*, 258(9): 1960–1967. https://doi.org/10.1016/j.foreco.2009.07.045

Roturier, S., Sundén, M. and Bergsten, U. 2011. Re-establishment rate of reindeer lichen species following conventional disc trenching and HuMinMix soil preparation in Pinus-lichen clear-cut stands: A survey study in northern Sweden. *Scandinavian Journal of Forest Research*, 26(2): 90–98. https://doi.org/10.1080/02827581.2010.528019

Roturier, S., Bäcklund, S., Sundén, M. and Bergsten, U. 2007. Influence of ground substrate on establishment of reindeer lichen after artificial dispersal. *Silva Fennica*, 41(2): 269–280.

Roturier, S., Ollier, S., Nutti, L.-E., Bergsten, U. and Winsa, H. 2017. Restoration of reindeer lichen pastures after forest fire in northern Sweden: Seven years of results. *Ecological Engineering*, 108: 143–151. https://doi.org/10.1016/j.ecoleng.2017.07.011

Sandström, C. and Widmark, C. 2007. Stakeholders' perception of consultations as tools for co-management: A case study of the forestry and reindeer herding sectors in northern Sweden. *Forest Policy and Economics*, 10: 25–35. https://doi.org/10.1016/j.forpol.2007.02.001

Sandström, P., Cory, N., Svensson, J., Hedenås, H., Jougda, L. and Borchert, N. 2016. On the decline of ground lichen forests in the Swedish boreal landscape: Implications for reindeer husbandry and sustainable forest management. *Ambio*, https://doi.org/10.1007/s13280–015-0759-0

Sandström, P., Granqvist Pahlén, T., Edenius, L., Tømmervik, H., Hagner, O., Hemberg, L., Olsson, H., Baer, K., Stenlund, T., Göran Brandt, L. and Egberth, M. 2003. Conflict resolution by participatory management: Remote sensing and GIS as tools for communicating land-use needs for reindeer herding in northern Sweden. *Ambio*, 32: 557–567. https://doi.org/10.1579/0044-7447-32.8.557

Wallenius, T. H. 2011. Major decline in fires in coniferous forests: Reconstructing the phenomenon and seeking for the cause. *Silva Fennica*, 45: 139–155. https://doi.org/10.14214/sf.36

Wenger, E. 1998. *Communities of Practice: Learning, Meaning, and Identity*. Cambridge: Cambridge University Press.

Young, T. P. 2000. Restoration ecology and conservation biology. *Biological Conservation*, 92: 73–83. https://doi.org/10.1016/S0006–3207(99)00057-9

Zackrisson, O. 1977. Influence of forest fires on the North Swedish boreal forest. *Oikos*, 29(1): 22–32. https://doi.org/10.2307/3543289

Part II

Indigenous Perspectives on Environmental Change

8

The Climate Agreements: What We Have Achieved and the Gaps That Remain

HINDOU OUMAROU IBRAHIM

I was elected Co-Chair of the International Indigenous Peoples' Forum on Climate Change for the first time in December 2014 at COP 20 in Lima, Peru (20th Meeting of the Conference of Parties to the United Nations Framework Convention on Climate Change) even though I neither asked for the position nor wanted it (Fig. 8.1). The Chair wasn't able to stay for the second week of negotiations, so he asked me to assist him in completing the work. As the appointment of Chairs takes into account the location of the COP, I was then asked to continue in this function at COP 21 in 2015 that was held in Paris, where I reside part of the year. Normally, my contribution would have ended there. But as COP 22 was to be held in Africa, at Marrakech in 2016, I was asked once more to continue for yet another term as I am African and speak French, English and Arabic. After COP 22, I insisted that I wanted to step down but no one else wanted

Figure 8.1 Hindou Oumarou at the International Conference *Resilience in a Time of Uncertainty: Indigenous Peoples and Climate Change* held at UNESCO during UNFCCC COP 21 (2015).

to shoulder the responsibility. So I was once more re-appointed Co-Chair, for the fourth and last time, at COP 23 in Bonn in November 2017. These three years were very intense with endless travels, meetings, conferences and difficult negotiations. Here I recall the events that took place at the COPs in Paris in 2015 and in Marrakech in 2016, where I contributed actively to the major advances secured for Indigenous peoples' rights at these events.

Indigenous Peoples and the Paris Agreement : UNFCCC COP 21 (Paris, 2015)

As Co-Chair of the Indigenous Peoples' Caucus, or more precisely the International Indigenous Peoples' Forum on Climate Change, I am responsible for coordinating Indigenous peoples' participation from our seven socio-cultural regions, namely, Africa; Asia; Latin America and the Caribbean; North America; the Pacific; the Arctic; and the Russian Federation and Central and Eastern Europe. Each of these seven regions has its own specificities, but in international negotiations, our collective efforts focus on a single issue: the international agreement and its impacts on our communities. This task is completely different from what I do at the national and local level for my country and my people.

Our actions have a dual focus: the technical negotiations and the political negotiations, the former laying the groundwork for decisions to be taken by the latter. Preparatory meetings had already been organized within each of the seven regions so as to prepare for technical negotiations at the Paris meeting held from 30 November to 11 December 2015. These outcomes were then brought together by a technical team that helped us prepare a joint policy statement in the form of a declaration. On the basis of this joint policy statement, we convened a meeting of the fourteen members from these regions (two from each region) in order to define our objectives for Paris.

Our Paris objectives provided the technical basis for two meetings with Member States: these dialogues between Indigenous peoples and States were held in Bonn at the UNFCCC secretariat and in Paris at UNESCO. We invited forty-one States that are 'friends of Indigenous peoples', not all of whom were able to attend. We presented our expectations for the Paris Agreement, explaining where we would like references to the rights of Indigenous peoples to be included, as well as mention of the traditional knowledge of Indigenous peoples.

When we arrived in Paris for the formal negotiations, it was relatively easy for me to liaise among my team, the French Presidency and the UNFCCC secretariat. The State–Indigenous dialogue sessions greatly facilitated our negotiation of the Paris Agreement, in which we managed to include five references to Indigenous peoples in the Agreement and in the Decisions. Of course, this process was also supported by numerous other bilateral and informal exchanges, as well as official meetings with heads of state and government, such as the then President of France, Francois Hollande, among others.

In the Agreement, we obtained a reference to Indigenous peoples' rights in the Preamble, and also in Article 7, paragraph 5, that refers to the knowledge of Indigenous peoples as a resource for climate change adaptation. These points in the Paris Agreement were also included in the Decision adopting the Agreement. There, Indigenous peoples' rights and participation are referred to in the Preamble, but there is also the very important Decision 135 that makes reference to an Indigenous Peoples and Local Communities Platform on climate adaptation and mitigation.

Overall, the Paris Agreement with its five references is a major step forward for us, especially because States recognize Indigenous peoples' knowledge up front in the operational part of the Agreement. Indigenous peoples' rights, on the other hand, only appear in the Preamble. Of course, the Preamble frames everything included in the articles. The problem, however, is that it is not necessarily binding, as it depends on the legal environment in which it is applied. In my opinion, it is deplorable to simply take on board the positive aspects of Indigenous peoples while leaving them out of the parts that are restrictive, that require States to respect us for who we are and what we do. That is the view I have on the Paris Agreement.

Implementation of the Paris Agreement

What is important for us is how the Paris Agreement is implemented. What measures will be taken by States to ensure respect for those rights that are the most important for us? When we consider the nationally determined contributions (NDCs) submitted by 176 States, only 24 countries include a mention of human rights. Seven of them state, 'Yes, they are important in the national context,' and fourteen state, 'They are important for implementation.'

Among these fourteen, there is my own country, Chad. This outcome is a result of the work that I was able to accomplish with my association at the national level and with the climate team that prepared the NDCs for Chad. We made it clear that the contribution could not be submitted without reference to respect for human rights, as it is the people who suffer impacts from a degrading environment. We also demanded that the NDCs include a reference to biodiversity. As a result, owing to our work upstream, we managed to include references to human rights in our national contribution.

Human rights were not even mentioned in 95 per cent of country submissions. How will they go about implementing the Paris Agreement? This is where there is a huge gap. Another huge gap relates to finance. The Agreement does not speak to the issue of financing the fight against climate change. There is only reference to financing the Decisions and even that is not well defined. So this was the central issue to be addressed at Marrakech (22nd Meeting of the Conference of Parties to the UNFCCC, 7–18 December 2016): defining the finance mechanism, accessibility, and also defining transparency. The Least Developed Countries, in Africa and elsewhere, insisted on re-negotiating these matters at Marrakech so that the Paris Agreement would be truly binding and oblige countries to respect the engagements made.

Funding Implementation of the Paris Agreement: Green Climate and Adaptation Funds

What are the implications for Indigenous peoples? First, there is the Green Climate Fund (GCF). For sure, it has set in place a series of environmental and social safeguards. At the same time, however, we have only one observer participating in meetings of the Board or other major events of the Fund. At the Marrakech COP, we requested that the GCF allow greater participation of Indigenous peoples in order to better represent our communities and priorities. We proposed the designation of a focal point specifically for Indigenous peoples within the Fund secretariat. We also requested the establishment of an informal consultative group that would monitor and eventually re-jig projects submitted by States: Are the proposed projects respecting Indigenous peoples' rights or are they in violation, in which case, we would oppose them. What safeguards are needed to ensure that projects are implemented in a manner that respects the rights of communities and ensures their effective participation so they also benefit? We met with the GCF secretariat to discuss these matters, and the head of the secretariat assured us that a focal point would be nominated to put such measures in place.

We also requested that the GCF adopt an Indigenous peoples' policy, like the policies adopted by the World Bank, United Nations Development Programme, Food and Agriculture Organization, International Fund for Agricultural Development, Global Environment Fund, etc. Indeed, all funding agencies and UN organizations, including the World Bank, have developed specific policies with respect to Indigenous peoples.[1]

Another aspect of implementing the Paris Agreement is the large-scale development projects already submitted to the GCF. These megaprojects are not at a scale that provides benefits to local populations. For the developed countries that provide funding, many of these projects would seem appropriate because they generate renewable energy. However, projects such as mega-dams pose major risks for biodiversity and displace large numbers of people from their homelands with a total disregard for human rights. When projects of this nature are being considered by the GCF, we requested that representatives of Indigenous peoples be invited to sit at the table with States to discuss the potential impacts on their communities. The Belo Monte Dam in Brazil is one such example. It is a classic case: financed by the banks and foreseen to provide renewable energy for 18 million people, or something like that. But it will displace Amazonian Indigenous communities and that is unacceptable for Indigenous peoples. It is their environment and way of life that should be protected first and foremost. This project generated a major conflict. The communities resisted and brought the case before the Inter-American Bank. The project was temporarily put on hold.

In response to another project presented to the GCF, Indigenous peoples submitted petitions because once more communities were to be displaced and human rights were not respected. This led to a protest at the Marrakech COP. In order to avoid these confrontations, we proposed the creation of an *ad hoc* committee to review project proposals upstream. If States come to justify their proposals, then we should also be invited to sit at the table with them to exchange views.

However, the GCF is not the only body supporting implementation of the Paris Agreement. There is also the Adaptation Fund. Even though it has a smaller funding pot, it could revolutionize access to financial support for Indigenous peoples. We are asking for direct access to the Adaptation Fund via a specific window for Indigenous peoples. This negotiation is also ongoing, but I believe that it could be accepted. Let us see how things evolve.

Indigenous Peoples and Negotiations during COP 22 at Marrakech in 2016

For the Marrakech COP, we also held preparatory meetings. As Co-Chair of the Indigenous peoples' caucus, I was responsible for informing everyone about the organization of these meetings. For the dialogue sessions between Indigenous peoples and the States, it is usually the host country that sends invitations to the others. In this case, it was Morocco that was to send out invitations to announce: We invite you to participate in a dialogue between Indigenous peoples and States on such and such topics. That's how it was done at COP 21 in Paris, where Laurence Tubiana, as chief negotiator for France, sent out the invitations and co-chaired all of the dialogues along with Peru and myself. At the Peru meeting (20th COP of the UNFCCC, 1–4 December 2014), it was the Minister of the Environment, Mr Manuel Pulgar-Vidal, who sent out the invitations and co-chaired with us.

So in preparation for Marrakech, I went directly to Rabat for three days in order to work on the invitation letter, the agenda and logistics, alongside their climate team, the Minister of Foreign Affairs and the Minister of the Environment. Once the invitation letter was completed, the Minister of the Environment sent it out to all of the States. An additional reason for my visit was to oversee the construction of the Indigenous peoples' pavilion, where people who do not have access to the negotiations can put up exhibitions and displays illustrating Indigenous peoples' actions to combat and adapt to climate change.

The dialogue meeting took place on 5 December 2016. Both the Chair of COP 22, Salaheddine Mezouar, and the Minister of the Environment, Hakima El Haite, were with us during the dialogue. We presented the Indigenous peoples' vision of how the Paris Agreement should be implemented: both the Preamble on the rights of Indigenous peoples, and Article 7, on the recognition of the knowledge of Indigenous peoples as a source of solutions for climate change.

We discussed with the States and came to an agreement on how to operationalize the platform for the exchange of traditional knowledge. The knowledge platform was actively supported by Bolivia and Ecuador. When I was in Rabat, I spoke to the Chief Negotiator Aziz Mekouar, and he told me, 'Don't worry, your concerns are already addressed by Bolivia. So what are you waiting for?' So I sent him the positions that we wanted to push forward. As we had already discussed our expectations with States during the dialogue meeting, it was much easier for us to make headway during the negotiations.

Once negotiations began, we had two things to follow: activities in the Indigenous peoples' pavilion in the green zone which is open to the public, and formal negotiations in the blue zone which is only accessible to representatives of States and select organizations. During negotiations, my usual pattern is to chair each day the coordination meeting of the Indigenous peoples' caucus from 9 to 10 AM. Each year, the secretariat of the UNFCCC provides us with a meeting room for this daily event that appears in the official UNFCCC agenda. I fix the daily agenda and chair our discussions. I invite State parties that are able to attend and, in return, we attend their meetings or meet bilaterally.

A decisive moment for us was the secretariat's request that we meet to discuss the Indigenous peoples and local communities knowledge platform. We participated in a brainstorming meeting where we contributed our ideas and they developed them into three questions: what do we mean by 'exchange'; what do we mean by 'knowledge'; and what is the nature of the 'platform': is it a website? As we had a clear idea of what was needed, we replied, 'No, that's not how we envision the platform being implemented. If the platform is a website or on Twitter or Facebook, then that already exists. We already have our own. But our knowledge is much more than that. Our knowledge is on the ground and hands on.' So they asked us to make a proposal. We drew up a three-page proposal outlining what we expected and submitted it for discussion at a meeting of the COP 22 presidency and the UNFCCC secretariat. My role was to compile all inputs and, once finalized, to send it to the secretariat and the Presidency.

I submitted our inputs and, based on our submission, they convened a negotiation. It was titled PC21/para 135 on a platform for the exchange of Indigenous peoples' knowledge.

'Speaking about Us, for Us, but without Us'

Then, the first negotiation began. We were informed just one day prior. We were not at all pleased that they did not inform us well in advance so that we could be better prepared. During our preparation meeting, we agreed among the members of the caucus that Indigenous peoples must absolutely be allowed to take the floor. We would not just be observers. We were determined that they were not going to speak about us, without us. I wrote to the COP 22 secretariat and called them by phone to say, 'We must absolutely take the floor.' Their reply was that under UNFCCC procedures, only State parties negotiate. Even though meetings are open to observers, observers are not allowed to take the floor. But we were adamant, 'That's out of the question because you will be talking about us, for us, but without us. That's out of the question!'

I made preparations. I went down to the green zone to pick up some white tape. We were in the meeting room and I said to all of my Indigenous colleagues,

You stay at the back of the room. Here's the tape. I will negotiate with them as soon as they come in. If they let us take the floor, then we will speak. We are going to request to speak first, before the States. If they refuse to allow us to take the floor, we will tape our mouths shut and we will stay in the room with them. In the time it takes them to call security, at least two or three media will have seen us ... because this is completely unacceptable.

We were all there early at the back of the room. The secretariat and the Presidency arrived, and the State parties started to enter the room. We went to see Ecuador, Bolivia and Guatemala directly to say to them, 'We would like to take the floor first, before all of you. We know that the secretariat is telling us that this is not according to procedure, but you cannot do this without us. We must take the floor before you do.'

I went to see the secretariat and the COP 22 Presidency to say, 'You did not reply to my message, but we must absolutely be the first to take the floor.' The Presidency told me, 'We understand completely. But the procedure prevents us from giving you the floor, except at the end after the States speak.' I said, 'That's out of the question.' So he said, 'Okay, hold on. I'll open the meeting and let's at least see what the States propose.' I replied, 'Yes, we have some States that will make a proposal.' After some introductory words, he opened the meeting. Ecuador took the floor to say, 'We believe that Indigenous peoples should speak before any of us take the floor.' Bolivia took the floor to say: 'We support this proposal.'

The European Union wanted to contest, but all of the other States took the floor one after the other to declare: 'Okay, we agree that they speak before we do.' So we took a place at the table and we spoke. We made our statement before the State parties. After that, the negotiations were open. When the negotiations began, Canada gave the floor to the Indigenous members of its delegation to speak, saying, 'We wish that Indigenous peoples speak for themselves. We are not going to speak in their place.' It was a very interesting moment. All of the States were a bit shocked, but they remained silent nonetheless. Nicaragua had an Indigenous member on their delegation. New Zealand also had an Indigenous delegate. They sat at the table with their delegations but did not take the floor. At the end of the discussions, I requested that the UN Special Rapporteur on the Rights of Indigenous Peoples also be given the floor. She was given the floor and addressed the meeting.

After this first negotiation was over, we revisited our proposal in order to include what the States had said. We came up with a number of critical points. We said, 'Yes, Indigenous participation is important, but it must be from the very beginning of the process. Secondly, what we have to say has to be taken seriously, and for that it must be in the framework of the SBSTA (Subsidiary Body for Scientific and Technological Advice) or the SBI (Subsidiary Body for Implementation), so that it will be treated as a specific issue for negotiation.' We stated, 'Submissions should be made on the nature of the platform and how it will function.' We also said that we want to participate but we want it to be a true participation with us at the table to negotiate with the States.

The Presidency and the secretariat then convened another negotiation. They drew up a few well-defined points that addressed our expectations in part but differed from what most other parties were anticipating. This time we found out only a few hours prior that a meeting had been convened. We came to the COP venue in the morning, looked at the day's agenda and noted, 'Okay, there is a negotiation for us but they didn't even inform us.' We were not at all pleased. I called the secretariat and also sent a message saying, 'At least you should invite us or keep us informed of what is going on. And are we going to be given the floor? How is the meeting going to be organized?' They replied, 'Everyone can intervene and you can take the floor when and as you want to.' This was a real step forward!

When we came to the meeting room, things had changed. We were all seated together. But they asked for the States to intervene and when they had presented their four-point proposition, they gave us five minutes to look them over. Along with my Indigenous colleagues, we had already agreed on our strategy. It was that I would intervene and no one else. I would speak in the name of everyone to denounce the process. We said that either it is a real negotiation or we stop everything. Then Bolivia intervened, followed by the European Union and Guatemala. Nicaragua came over to where I was seated to say, 'Come with us. You can sit behind our nameplate and speak in our name.' When Nicaragua was given the floor, the delegate said, 'Thank you. I support what Bolivia has said because we need time to analyse the proposal and come back with our response.' And he added, 'I give the floor to my colleague, Hindou, who is here and will speak in the name of Nicaragua.' At that moment, it was finally clear to all the States that we could speak on an equal footing with them. I stated,

We agree with Bolivia. We cannot accept these four points. They are new to us. We need to go back to the Indigenous caucus to analyse this proposal and make our comments before coming back to you. So if you can convene another negotiation that's fine, but if you can't convene another meeting, we do not agree with these points.

The meeting was adjourned. At that point, we began to make urgent calls to the secretariat, the Presidency and others because no other negotiation was foreseen. It was already 16 December and on the 18th, the COP was to close. So we began a series of bilateral meetings, and I saw the women ministers on the morning of the 17th at a breakfast meeting on human rights organized by Mary Robinson, the UN High Commissioner for Human Rights. There, I spoke to the Minister from Luxembourg, Carole Dieschbourg, with whom I had become acquainted during her tenure as President of the COP 21 for the European Union. I explained the situation and that the European Union wanted to block our negotiations. Right away she organized a meeting with four other EU ministers in order to discuss the situation. Bolivia, Ecuador, Guatemala and others came to make a proposal in relation to what we had accomplished. We said we wanted the platform to be under SBSTA and that we would co-chair with the SBSTA chair. Also, we agreed to accept submissions from State parties along with submissions from other actors. We came to an agreement, and the Indigenous caucus the next morning was attended by Nicaragua, Bolivia and Australia. We then went to see the United States, Canada and the EU. We all came to agree on a text. We shared the text with everyone and sent it to the COP 22 Presidency.

A Major Victory for Indigenous Peoples

The text was adopted on 18 December at 2:45 AM under the COP 22 presidency. There were no objections because once more States took the floor to say, 'We are pleased with the text and we support its adoption.' In brief, it said that the platform had already been created and now it had to be operationalized. Second, it would be under the SBSTA and negotiated by the SBSTA chair: in particular at SBSTA 46 that was to take place in Bonn in 2017. This negotiation was to be co-chaired by

Indigenous peoples and the chair of SBSTA. Third, all States were invited to submit, by 31 March 2017, their positions with respect to the operationalization of the platform. The fourth point was that the secretariat was to prepare a report summarizing all of the submissions received. The final point was to hold a meeting between Indigenous peoples and the States during SBSTA 47, and that it would be clearly included as an agenda item for that meeting.

For Indigenous peoples, this was really a major victory: and also one for me, who stayed at the negotiations until 3 AM. It was a massive victory because for the first time in the history of the UNFCCC, the participation of Indigenous peoples was accepted: giving us the floor prior to the States; addressing the meeting from behind the plates of State parties; and formally adopting a decision whereby Indigenous peoples co-chair alongside the SBSTA chair. Nothing like this had ever happened before. After many long years of negotiation, it was a complete revolution. So there you have it. That's it!

Some Final Thoughts

Being Co-Chair of the Indigenous peoples' caucus is not an easy task. I had to forego many projects for my organization at the national level, and I never stopped travelling. But most of all, I never stopped working before and after each COP. For COP 21, during the last week, I spent every night in the negotiations and barely slept. I stayed up until 5 AM and started again the next morning at 7 or 8 AM. I was the last to leave the meeting and the first to arrive in the morning to prepare the agenda and all of the rest. Normally, there are three Co-Chairs, but my other Co-Chairs did not have the chance to stay the full two weeks of the COP. I don't regret anything because I was able to witness and experience all kinds of things during COP 21, and during COP 22 as well. And after all, I am proud of the outcome and proud to serve my relatives, Indigenous peoples around the world. But of course, after the COP was over, I was exhausted and very sick.

Note

1 The GCF Indigenous Peoples Policy was adopted by the GCF Board in 2018.

9

Reinforcing Traditional Knowledge in the City: Canoe Building and Navigation in the Changing Pacific

TIKOIDELAIMAKOTU TUIMOCE FULUNA

I come from Fiji, whose capital city, Suva, has a population of approximately 90,000 people. My family and others like mine, who are from remote smaller islands and moved to Suva hoping for a better life, make up about 15-20 per cent of this number. With this rural to urban drift, people lose links to their traditions and knowledge, especially as children grow up in an urban setting, lacking exposure to the stories, traditional skills and wisdom of elders who remain in the village. Nevertheless, we still strive daily to show the nation and also the world

Figure 9.1 The author in front of the *masi* cloth made by his Auntie Olivia.
Photo courtesy Tikoidelaimakotu Tuimoce Fuluna

that we are proud of who we are and how important it is for us to keep our traditional knowledge intact in this modernizing world. Thanks to this knowledge, we can contribute towards preserving our ocean.

The Origins of Our Ancestors

My family is originally from Moce Island, part of the Lau Island group, which comprises about 25 inhabited islands 275 km east of the main islands of Fiji. The people from my island of Moce, as well as those from several nearby islands including Kabara, Ogea and Fulaga, share in common the skills of building, sailing and navigating traditional double-hulled sailing canoes in order to carry out vital fishing and inter-island trade. We have a long tradition of canoe building that goes back thousands of years and is kept alive through the stories of our elders and the knowledge, wisdom and skills that they pass along to the next generation.

Moce Island is a high and fertile island with lush vegetation and good farmlands. The other three are volcanic islands with about 15 per cent of their surface having soil and the rest composed largely of volcanic rocks. For this reason, root crops including taro, sweet potatoes and yams, are very scarce on these islands. However, they do have an abundance of large native hardwood trees known as *vesi* (*Intsia bijuga*). My people are very skilled in the arts of building canoes and houses, and carving statues, *kava* bowls and *lali* (Fijian drums). Moce Island does not have as many trees as our neighbouring islands. In order to build a sailing canoe, the common practice within my community is to go to Ogea, Fulaga or Kabara, build a canoe, sail it home to Moce more than 50 km away to get supplies and farm produce, then sail back to the island that hosted the canoe building team and drop off the supplies as a barter payment for the tree and the hospitality. After the payment, the team returns home to Moce with their new canoe.

It was during one of those trips that my grandfather met my grandmother, who was from the chiefly craftsman's family of Ogea. My grandfather was a master navigator and master sailor. He was also a church minister who sailed on a traditional canoe to carry out his tasks and take the gospel to the islands. My grandfather had fourteen children and he passed on his knowledge of canoe building, sailing and navigating down to my father and his brothers, and they passed the tradition on to me (Fig. 9.2).

Seeking Opportunity in Suva

My father sailed to Suva in 1993 to start a new life and provide us with a better future. He had visited Suva to work and saw how different it was from village life

Figure 9.2 Me as a child, yearning to follow my ancestors as an ocean voyager.
Photo courtesy Tikoidelaimakotu Tuimoce Fuluna

and understood that for his children's future, moving to the city was important. The main challenge on the remote home islands is the lack of schooling. If we stayed on in the village, we would not receive a formal education. So, he sailed to Suva with my grandfather, hoping to start a business of traditional canoe cruises in the city. All was going well and there were about five or six canoes on the main island ready to start the business. But tragedy struck on 23 September 1994 when my father and a young cousin decided to sail to Suva on their newly built *camakau* (outrigger sailing canoe). They were lost at sea and their bodies were never found.

We were then raised by my mother in Suva. As the years went by, the land from where my father had planned to launch the business was filled with relatives from the island. Lacking enough traditional canoes and my father's presence, the business plan was shattered. My father's older brothers decided to keep and maintain at least some of the canoes and taught us how to sail and build them, keeping an ember of our traditions aglow.

Retaining Traditional Knowledge in Korova

The land on the shore in Suva that was supposed to be for the business is now named Korova. In my island of Moce, the only two villages are named Nasau and

Korotolu. *Korotolu* means 'third village' and *Korova* means 'fourth village'. Even though we were far from Moce, the elders kept the spirit of our ancestors alive in part by calling our small parcel of land the 'fourth village'.

Life in Suva has changed for most of the people who immigrated from smaller Fijian islands, but the people of Korova still practise the skills that the ancestors passed down through the generations: building, sailing and racing traditional canoes. Just a ten-minute drive from the city, and in the same neighbourhood as some large institutions with cars and buses on the busy streets, the men and women of Korova strive daily to show the nation, and also the world, that we are proud of who we are. Every day, we demonstrate the importance of keeping our traditional knowledge intact in this modernizing world.

During a weekend back at the village, the old and young men tend to their canoes, while our women make *masi*, which is the art of producing designs out of bark cloth. The *masi* plant takes approximately nine months to mature. When it is harvested, the inner bark is peeled off and then soaked in sea water to make it soft. After a week of soaking, the bark is laid on a piece of wood and beaten with a traditional tool known as an *ike* that is 30–40 cm long with 5–7 cm square sides and a rounded handle. The beating continues until a suitable width is obtained. These pieces are then joined together to attain the desired size of about 2 m by 0.5 m. After that process, it is laid out to dry in the sun. Once dry, the edges are trimmed nicely, and the sheet is printed using traditional dyes. These pieces can also be glued together to make large cloths that may be several meters long. Traditionally, the glue on the islands was made from the sap of trees such as the breadfruit tree. But it is now difficult to source that glue so we normally use common wood glue purchased in the city. These traditional *masi* cloths are used for weddings, birthdays and important ceremonies. It symbolizes our history and identifies where we come from, which is why I chose a *masi* cloth created by my Auntie Olivia as a backdrop for my photo (Fig. 9.1).

While the women of the village dedicate themselves to their traditional tasks, the men work on the canoes that we still use for fishing. Young boys open their ears when the older generations explain things about canoes or narrate how our ancestors used to travel from one island to another in their bigger and faster canoes. The boys open their eyes as the elders demonstrate how to carve and join the pieces of the canoe. They are fascinated by how the elders can read weather patterns by looking at the sky and, at night, by the stories they tell of how to navigate by the stars. Today, compared to my father's time, our elders say we are just 'building toys' since the small canoes cannot compare to the large voyaging canoes that our people once built. At Korova, we are confronted with a lack of materials together with insufficient knowledge about building traditional canoes. Still, we enjoy teaching the younger generation about sailing

Figure 9.3 My uncle Rogo teaching his son how to chip the hull with an adze.
Photo courtesy of Kerry Donovan

and building canoes and they remain eager to acquire traditional knowledge (Figs. 9.3 and 9.4).

That's the wonderful benefit of living together as a Moce community in Korova, in the middle of Suva, where we can keep our family heritage and traditions intact. As the elder generations share their stories about sailing canoes at the age of 12, 13 or 14, it makes our youngest generation think, 'Oh, if the older generations can do it, why can't we?' So, they start practising by building small model canoes named *bakanawa*. As we build our ocean-going canoes on land at the foreshore, we see the younger children playing nearby in the sea with their *bakanawa* canoes, challenging each other to races and dreaming of building a large canoe one day.

Traditional Canoe Races Help Keep Our Skills Alive

Pacific Blue Foundation, an organization with a focus on cultural conservation, has worked closely with my village as well as with communities from Fulaga and Ogea to fan the ember of our canoe building and sailing skills through inter-

Figure 9.4 Semiti, the eldest of my father's siblings, with me during construction
of a *drua*.
Photo courtesy of Kerry Donavan

community races that we have kept alive. In 2010, Pacific Blue Foundation
initiated an annual race of traditional sailing canoes, 'Veitau Waqa – the boat lives'
in Fiji's capital, Suva (Fig. 9.5). There were only four canoes in the first event but
thanks to the efforts of the community, funding from iTaukei Lands Trust Board
and Pacific Blue Foundation, by 2014 the event had ten canoes. This has brought
national and international awareness to our canoe building and sailing traditions.

In addition to this race of larger canoes, our children challenge each other with
their smaller model canoes called *bakanawa* (Fig. 9.6). As the elders are preparing
their boats for race day, the children pool together a small amount of money from
each participant and the winner takes all. This type of activity really enhances our
knowledge, improves our skills and at the same time helps us pass on to the
children the spirit of who we are and where we come from. By the time the annual
'Veitau Waqa – the boat lives' race day rolls around, more than 100 children from
many communities arrive with their *bakanawa* for a challenge race!

We build large canoes based on knowledge passed on orally from our elders.
As it takes many months, starting with selecting trees in the forest to finishing the

Figure 9.5 'Veitau Waqa – the boat lives' – the annual races at Suva Harbour that were initiated by Pacific Blue Foundation as part of efforts to keep traditional cultures alive. By 2014, there were ten smaller *camakau* and two larger *drua* available for the 'Veitau Waqa' event. [A black and white version of this figure will appear in some formats. For the colour version, please refer to the plate section.]
Photo courtesy of Aaron March

Figure 9.6 The children challenge each other in the *bakanawa* race during a mid-day break at the annual 'Veitau Waqa-the boat lives' event. [A black and white version of this figure will appear in some formats. For the colour version, please refer to the plate section.]
Photo courtesy Aaron March

canoe, the younger generations watch us and become curious. They ask many questions, 'What is this? What is that? How does this work? How does that work? How do you tie the knot to join this piece to that one?' We are happy to share with them the knowledge we have received from our elders, knowing that one day we will be gone and these children will be the custodians of our ancient traditions and cultural identity. It is gratifying to know that the youth are eager to get this knowledge from us and we take pride in passing on what we learnt from our elders.

The use of modern fibreglass boats powered by fossil fuels for fishing, and government ferries for local as well as more distant inter-island transport, has resulted in a decline in canoe construction throughout the Pacific Islands. Now the traditional smaller *camakau* canoes with their one hull and small outrigger that were common when I was a child no longer exist on our home island of Moce. Also, they have not constructed a larger *drua* on the remote islands since four decades. *Drua* means 'twins' because the boat design has two large hulls. To offset this decline in traditional knowledge on the home island, in the past decade we have built at Korova both *camakau* and *drua*. We pass this knowledge to those in the next generation who are willing to learn, knowing that if we lose this, we lose our culture, our heritage and our identity.

National and International Recognition

After we began the 'Veitau Waqa – the boat lives' races in 2010 with the assistance of Pacific Blue Foundation, Fijians became more aware of our efforts through local press, television, radio and documentaries. I was fortunate during this time to be heavily involved in building traditional canoes and maintaining them with a few of my older uncles. Eventually, the word spread internationally and in 2012, the *New York Times* ran an article entitled 'Looking Forward, Fiji Turns to Its Canoeing Past' by Ginanne Brownell (*New York Times*, 3 February 2012).

In 2014–15, Joji Misaele of the Fiji National University collaborated with Steven Hooper of the Sainsbury Research Unit at the University of East Anglia (UK), on a project funded by the German Government to build a *drua* named 'Adi Yeta'. This 8 m long *drua* sailed in the 2015 Veitau Waqa races and was displayed at the 'Fiji' exhibition in England. Along with the Fiji Army band, she featured at Her Majesty Queen Elizabeth II's ninetieth birthday pageant at Windsor Castle. 'Adi Yeta' is now on permanent display in the 'Pacific Encounters' Gallery at the National Maritime Museum in Greenwich, London.

Later, the Walt Disney Animation Studio team spent time in our village, and their animation of the canoes in the film 'Moana' was based on the boats we had built. To show their gratitude, Disney financially supported our construction of a *drua* in 2015 through a donation administered by the Pacific Blue Foundation.

Men in my village went to the forest near Suva, harvested the trees, carved and chipped the logs, then transported them with a large truck to Korova. We built it traditionally by joining all parts with *magimagi* or coconut husk rope, in the traditional way. Our elders named the *drua* 'Heart of Te Fiti' to honour the fact that the boat was sponsored by Disney, whose film Moana was a quest to find and restore the stolen 'Heart of Te Fiti'. a gemstone that gave the goddess the power to bring life to the oceans (Fig. 9.7).

Trans-Pacific Ocean Voyaging Using the Stars, Sun and Wind

From 2012 to 2014, I was also fortunate enough to be part of a voyage involving seven canoes from different Pacific Island countries: one canoe each from Fiji, Hawaii, Samoa, Tahiti and Tonga, and two from Aotearoa (New Zealand). This initiative was sponsored by The Okeanos Foundation for the Sea. The main aim was to carry our message of hope and show how important the oceans are to us, while appealing to humanity to not pollute the ocean and air as it is not necessary to burn fossil fuel to travel thousands of kilometres around the world. We did not use motors and accomplished the trans-Pacific voyage navigating with the stars and the sun while using the natural resource of wind power. We showed the world that we can do it and reconnected with the spirit of our ocean-voyaging ancestors.

I sailed on the Fijian canoe *Uto ni Yalo*, meaning 'heart of the spirit'. Sailing on this big canoe for two years allowed me to practise and become proficient at what I had learned in my early years from my forefathers. They used to speak about the wind, weather patterns, stars and other details needed to navigate the canoes, and I had once followed them on a short trip when I was young. But when I went on a two-year voyage in the deep open ocean, I developed much greater respect for what my elders had been communicating to us: how to read the wind and the stars, when the sun rises and sets, when different constellations rise and set, and how the clouds and the winds can predict the weather and the state of the sea. From these details the ancient knowledge allowed them to know their way. These deeply held traditions and knowledge, passed on through the generations for several thousands of years, also inspired the Lin-Manuel Miranda song, 'We know the way' that was featured in the film 'Moana',

> *We are explorers reading every sign*
> *We tell the stories of our elders in a never-ending chain*

It is our culture to voyage on the ocean winds using what we have learned from our ancestors in a never-ending chain of knowledge. We do not need to burn fossil fuels to be able to travel. We do not have to rely on engines to take us fishing or to gather seashells when we have our sailing canoes. Instead, we use the wind to

(a)

(b)

Figure 9.7 Two recently built *drua*: (a) with my uncle Semiti Cama on the maiden voyage of the *drua* 'Heart of Te Fiti' built by the Korova community; and (b) the 'Adi Yeta' (foreground) as the race begins at the 2015 edition of 'Veitau Waqa: the boat lives'. This *drua* is currently on display at the National Maritime Museum in Greenwich, London.
Photo (a) courtesy of Sefano Katz; Photo (b) courtesy of Roko Josefa Cinavilakeba

power our traditional canoes to gather our food. We have always relied on the sea for our survival, and we always try our best not to disturb the sea and the land as it has fed us and kept us alive. This is how our forefathers sailed from island to island, and it reminds us of the importance of Indigenous knowledge as a response to climate change.

We Are Ocean People and We Need Our Connection to the Sea

My island, together with the rest of the nation of Fiji, does not contribute much to climate change but we are victims of this uncontrolled and unplanned crisis that has forced many villages in Fiji to relocate during the past few years. After the severe Tropical Cyclone (TC) Winston battered Fiji in 2016, the government decided that villages which were decimated by massive waves and storm surge must be relocated to higher ground. The same happened during several large tropical cyclones in the past decade. Some of the villagers have asked the government to relocate their village because of fear that another cyclone might hit. TC Winston was a catastrophic cyclone; many houses were washed away, and this continues to be a growing threat as the strength of tropical cyclones is increasing with the warming of the oceans caused by climate change.

Even though my house at Korova is a mere five metres from the shoreline, luckily none of our homes were flooded or damaged by wind as the huge cyclone TC Winston passed farther north. However, our small piece of land has been eroding every year because of sea level rise and constant wave action resulting from the prevailing trade winds. Our land measured 10,000 m^2 back in the 1990s and now, twenty-six years later, it is down to only 5,000 m^2. We have managed to stay in the same place, but the future is uncertain. My village, Korova, has always had a close relationship with the sea. As we are surrounded by mangroves, we are always reminded not to disturb our surroundings and also to respect the ocean, as it has fed us and been our highway for millennia. We plan to plant more mangroves between our houses and the sea to help as a barrier. We believe in practising what our forefathers did, by respecting the ocean and the land, and enhancing our traditional knowledge, which includes maintaining a good relationship with the environment. In this way, we hope to keep our small village of Korova safe, as it has been kept for the past twenty-five years.

On my family island of Moce, 275 km east of Suva, sea level rise and the threat of more powerful tropical cyclones are serious concerns. But there is plenty of higher land available there, so in the coming decades we expect people to relocate further inland. It is one of the challenges traditional coastal villages will face in Fiji. If villagers are forced to relocate to higher ground, it would negatively impact traditional canoe building. The canoes are very heavy and must be built by the shore. Although I grew up in the Korova community in the city of Suva, we are still ocean people from small islands. If we are forced to move away from our small seaside parcel of Korova to the interior, it would be very difficult for us to build canoes. Our close community from Moce would be in disarray, and we could become more disoriented by the modern urban reality. By disconnecting our family ties, our culture can also be disconnected. Right now, there are only two elder men and a few elder women in our community who are living the traditional way, asking the younger generation to follow them. If we move from Korova, it will be a very big problem for us in terms of losing our skills. We will not be able to practise this tradition of canoe building in the interior, away from the seashore, and the dispersal of our community around the modern urban city will prevent the traditional education that passes knowledge from elders through adults to children.

The Challenge of Keeping Our Traditions Alive

The knowledge of building traditional sailing canoes, seafaring and navigating was passed down from my ancestors because of the vast ocean domain from which they came. What is now known as the Fiji Republic comprises more than seventy-five islands spread across more than 500 km east to west and 350 km north to south. Before the Europeans arrived, this vast expanse of ocean was linked by the canoes our ancestors built, allowing this large ocean region to share common traditions, language and culture. Therefore, the skills of our ancestors, who lived in a small cluster of islands of what is now the easternmost part of the nation of Fiji, are honoured and revered for their foundational role in contributing to Fijian culture and traditions as ocean voyaging people. In fact, all Fijian children are taught in school the story of the ancient canoes that first arrived in Fiji. Elders in all villages, not just our island cluster where the canoes were built, tell stories of the canoe as a fundamental part of our traditional culture. Indeed, the skill of building, sailing and navigating traditional canoes is one of the most essential elements of our cultural heritage.

Unfortunately, with the globalization of world economies and reliance by present-day island dwellers on fibreglass boats with engines driven by fossil fuels, there are very few Fijians remaining who continue to keep canoe building and sailing alive. We also have challenges with the changing climate. We see changes in the weather patterns now compared to those our forefathers used to see. In the old days, the wind would come from the east and southeast for about two weeks before turning southerly, allowing them to take advantage of this and plan longer voyages from their home island and back. They would look at the horizon in the evening and say, 'Oh it's southeast, so we can prepare our journey.' But, now it is different. The wind changes so suddenly. We can't really predict the weather as we did before. We can't be bold enough to say, 'The wind is going south, so tomorrow it will go from the west, probably southwest.' You can wake up the next morning, and the wind is northeast when you were not expecting that. So, it makes it hard to predict or navigate without modern technology.

To make new boats at Korova, it costs too much to bring trees from our island by ferry, so we look for other trees from the forest near Suva. Even though they are not the native tree that we prefer, they are sufficient for us to continue to carry on the canoe building traditions and with canoes we can still fish, race and navigate. The ember of our ancestral knowledge, passed down for thousands of years, continues to burn in our hearts. We are committed to passing our knowledge along to the next generation so that this never-ending chain of our culture continues into the future.

Acknowledgements

Contributions by the staff of Pacific Blue Foundation, including Roko Sau Josefa Cinavilakeba, Kerry Donovan and B. Greg Mitchell were essential for the re-establishment of canoe building by my Korova community. Dr Mitchell and Mr Donovan provided considerable assistance with preparation of the manuscript and the curation of the images.

10

Reindeer Herding in a Time of Growing Adversity

ANDERS HENRIKSEN BONGO

I was born in 1939 in a Sámi reindeer herding family and was a reindeer herder in Kautokeino (Norway) with my father and my brother until I was thirty-five years

Figure 10.1 Anders Bongo with Marie Roué on the front porch of his summer cabin at Aiseroaivvi, Norway (2017).
Photo by D. Nakashima

old. In the 1970s, I took the decision to stop herding (Fig. 10.1). There had been a number of difficult years with poor weather conditions, I had health problems, and my wife and I were interested in trying a new way of life that might be better for our children. We sold all our reindeer and took up new professions. I first worked a few years for the regional reindeer management office, before taking up a job with a company that fabricates tents for herders. Eventually, I bought this company, and it is now owned by my son. Today, I am seventy-seven years old and have lived in Kautokeino all my life, where I have family and friends who are, for the most part, reindeer herders. I will try and describe changes in reindeer herding practices that I have observed over the last few decades.

In the old days, there were smaller herds and not so much grazing pressure. Today, there are about three times more reindeer than there were in the 1960s and 1970s, but the area of pasture has remained largely the same with significant erosion from urban expansion and industrial development. People need bigger herds now because they need more money for snow scooters and cars. They need 500–600 animals just to make a living, and even then, the herder's spouse needs to take up a wage-earning job as well. Nowadays, there are around 100,000 reindeer in the winter area, while back then there were only about 30,000. And this is one of the reasons why problems occur.

Earlier, the herds were fairly well dispersed, but now they are so close to each other that they intermingle easily. When herds meet, they mix. When you have a small herd, you can easily keep an eye on it. If the herds intermingle – and they often do – it is a lot of work and takes days to separate them. Often, you can't do it immediately, so the herders have to work together for weeks, even months. Herders prefer to have control over their own herd because it is their money running around.

Today, the number of reindeer in Kautokeino has to be reduced. Some herders have 1,500 or 2,000 animals and that is too much. They need so much space. The government has come up with a new rule: each herder must slaughter a portion of their herd. The herders are not happy about this, even though they understand that there are too many animals. They have discussions about which herders will downsize their herds. This is a complicated question. Herders that have small herds of 200 reindeer can't cut down much because with fewer reindeer they will not be able to make ends meet. They already have so few to begin with. And the owners of big herds say, 'Okay, we have worked so hard for so many years to build up our herds, why should we be punished for that?' This is like the familiar discussion in everyday life, 'Who will get taxed the most? Is it the rich or the poor?'

In one district, for example, there was a young herder who had 150 reindeer. It's normal to start with just a few reindeer when you are young. The state told him to keep only 75. He refused. He sued the state in the fall of 2016, and he won!

However, the state appealed the decision. He warned that he would not give up, that he would take it to the highest court. He said that if he loses the appeal, they can take all his reindeer. For now, though, he has to pay his lawyer.[1]

There are so many people in reindeer herding and so many reindeer. That's the problem. For instance, in my former summer district, Aiseroaivvi, there are fourteen herders, now termed 'operation managers' by the government, and about 7,000 reindeer, which means an average of 500 reindeer per person or family. The authorities would like to keep it down to only 4,900 reindeer in Aiseroaivvi. When you do the maths, with fourteen operation units, that would mean an average of about 350 reindeer per unit, which is not viable. It takes about 500 reindeer just to make a living these days. That means some people would need to give up herding. Herders must find a solution among themselves, perhaps by first asking those who have 1,000 animals to cut down? For now, the authorities are asking that everyone downsizes their herd by 25 per cent. The one who has 1,500 reindeer would still have over 1,000 after putting down 25 or 30 per cent. The herder who has 300 reindeer, on the other hand, would only have 225 left, or even 210 with 30 per cent fewer! And that is too few to live off.

From 1980 onwards, herders were offered compensation of 400,000 Norwegian crowns[2] from the State if they agreed to stop herding reindeer. It is not easy for reindeer herders to abandon reindeer because herders don't have the formal education necessary to take up an alternative career. Nonetheless, in my summer *siida*[3] in Aiseroaivvi alone, ten families have given up herding since 1973. My family and I did not receive any compensation since we decided to abandon herding before these measures were adopted. Actually, several of the families who received these subsidies have not really abandoned herding or have returned to it. For instance, the son of the herder who relinquished herding would take it up again. And obviously the State had not thought about that. That is why these costly and inefficient measures have been abandoned. But today, there is talk about bringing them back into effect, in a much more specific way: all members of a family would have to give up herding and would not be allowed to resume later.

The Autumn and Spring Pastures

Earlier, there was better cooperation among herders, but a lot has changed in the last few years. There is no space to be flexible anymore. In summer, herds have their own place along the coast and in winter they go inland. Each herd has its own area from old times. In the summertime, as the reindeer are dispersed all along the coast, there is no problem. But during winter, they all gather in one small area and conditions are much more difficult. In between, they have the autumn pasture where they stop over before they go to the wintering areas. Earlier, there was what we called in Norwegian, *tommelfinger reglen*, a 'rule of thumb' – it wasn't written, but everybody respected it. You had to be out of the autumn area by the end of October. Now herders prefer to keep their herds longer on these autumn pastures because the Norwegian State has decided that they are shared pastures. As it is a commons that everyone uses, it is in each person's interest to graze as much as possible here in the intermediary autumn pastures. That way, they will not put too much pressure on their own winter pasture.

Climate Change in the Tundra

This problem has been exacerbated during the last ten years, as we've seen the cold weather come much later. While back in the 1970s we had snow in late September and often minus 40°C before Christmas, these last few years, the first snow has fallen only in late November, and very little snow at that. November and December used to be the best months for herding. It was cold and there were weeks when the temperature dropped to minus 30°C. Early snow and the early onset of cold weather were common. Now the temperature rarely drops below minus 20°C. Of course, we have some days when it is minus 30°C, but it is much rarer. Climate change! It is about 10°C warmer than it used to be when I was herding reindeer. Owing to the mild weather, the rivers freeze very late, which prevents the reindeer from crossing the river and going to their winter area, so they stay much longer in the intermediate area. They have to wait so long in the autumn, and also in the spring when they move the other way, that these intermediary pastures, where previously reindeer would usually just pass through during their migration, get eaten down. The reindeer eat and trample the lichen and it ruins the pastures in some places. This year (2016), by 12 December when all the herders were supposed to have reached their winter pasture, only one of the families was in the right place, in the winter place. All the others were still on their way between Jesjavre and Sjusjavre. So, of course, there is much pressure on this area.

Dividing the Winter Territories

Traditionally in Kautokeino, each herder or each family group used a specific location for their herd's winter pastures with nothing but common knowledge to guide them. Everybody knew where everybody else was and there was a lot more flexibility. At that time, I was a reindeer herder myself. The herds were smaller, so if one winter area received a lot of snow, or your herd could not find enough to eat there, then you had such good cooperation with your neighbour that you could move to his area. Often, there was perhaps just one kilometre between us. Sometimes, when I was herding, I could see my herd here and the neighbour's herd over there. We had much better cooperation then, everybody accepted that you had to move your herd and there was more understanding that some areas could have better snow and grazing possibilities. You had to communicate with each other. We didn't put down much on paper; it just worked out.

Many changes have taken place in the last few years. As herders acquired more reindeer, they began to have more conflicts. There were problems with the intermingling of herds, and it was not as easy as before to cooperate. So, the government decided that reindeer herders need borders in order that each one could say, 'Okay, this is your area. You can be there.' Otherwise, herders would be discussing all the time 'Who has the right to this place?' It is a conventional way of thinking; that you need borders everywhere; you need rules, and you need to put them on paper.

They have drawn borders in the winter area so that everybody knows, 'Inside this line, that's my area.' Earlier, it was in your mind, but nowadays you have it on paper that this area is your area. If you go into your neighbour's area, then the neighbour almost calls the police. Some people have even sued each other and taken them to court. The main problem is that they have such big herds that one cannot say anymore, 'Okay, you have bad snow conditions there, you can come over to my area.' They cannot do it because it is filled with their own reindeer. So, you haven't got the ability to be flexible anymore.

Losing Tundra to the Bushes: Or the Decline of Lichen

Thirty years ago, our pastures were mainly open tundra, or *duottar*, as we call it in our Sámi language. Now there are trees all over the place. That is another big problem caused by climate change; bushes grow instead of lichen. Lappoluobal is a place I know well. It was here, fifty years ago, that we would stop for the night and let the draft and pack reindeer (Sámi: *heargi*) graze because the lichen was plentiful and very high (about 2–3 fingers high, i.e., 3–5 cm). Now it is all bushes there, *muorra*, because of climate change. The bushes grow so tightly together that you can hardly see through them. I used to be able to see the church across the road from my little cabin but now I can't. I can't even see the path anymore because of these high bushes.

Before the 1980s, there were hardly any trees, just a little shrub where you could tie the reindeer. At that time, it was difficult to find a place to tie your reindeer as there were so few bushes. When you have a reindeer and want to tie it to a tree for the night, you need a tree without any others around it. This is because the reindeer needs enough space to be able to graze around the tree without entangling its leash on other trees or bushes. To have access to enough food, the rope should be at least 5 m long. So, you would need to find a tree without any others around in a 5-m radius. Today, as we can see, the bushes are too dense, it is like a forest, and it would be very hard to find a place to safely tie a reindeer for the night.

In this area, when we left a reindeer overnight, we tied it to a tree and came back the next morning. They could stay around 10–12 hours at the same place and graze around the tree. Then you had to move them and tie them to another tree so that they had a fresh patch to graze on. Nowadays, there is so little lichen left that within a maximum of two hours at the same place, they would have eaten everything available. Here, the reindeer can no longer survive. You have to give them hay or pellets to eat and that is what the herders do today.

There used to be small open spaces everywhere and good pasture because there is little snow in these places. Now in the open areas, we have these bushes growing, destroying the *jeagil* (Sámi term for reindeer lichen) and trapping a lot of snow there. So, it is more difficult for the reindeer to reach the lichen and it is like that all over the tundra. They're losing these areas to the bushes and that is making a big difference. Now it is impossible for the reindeer to graze there. There are more and more trees, which are taking over areas of pasture. The reindeer can eat crowberry bushes when they are hungry, but they prefer the lichen. When they are really hungry, they can also eat small bushes of birch in summer and in winter.

In the old days, there was lichen everywhere but now there is nothing because of climate change and also because too many reindeer have eaten it all. It is not only climate change; it can also happen when the summer is too dry, when there is no rain and the lichen cannot grow. Grass or trees grow fast but not lichen, which can take 100 years to grow, if I am not mistaken. Today, it is black earth, with nothing growing in it. I cannot even understand how the reindeer manage to eat by pulling out twigs of half a centimetre here and there. A catastrophe is waiting to happen. I think in ten to twenty years, reindeer won't be able to survive anymore.

Many people say, 'Oh, there are too many reindeer and that is the problem. They eat up all the *jeagil*.' But it is not only the reindeer that destroy the lichen; it's also the weather. Perhaps pollution is also a factor, what they call acid rain. Then the lichen becomes so dry that it dies out on its own. In Kautokeino, the climate is normally very dry and when it gets hotter, there is a chance of them drying out completely and not growing back. In recent years, it has been drier. If they don't do something to stop it, the lichen won't persist. And the *jeagil* takes a long time to grow back.

A New Problem in Autumn: Ptarmigan Hunters

Another problem nowadays comes from the ptarmigan (*Lagopus lagopus*), a grouse that lives in the tundra. I read that one fifth of the population of ptarmigan in the whole of Norway are found near Kautokeino, so it is the best area for ptarmigan hunting in the country. The hunting season starts on 10 September.

Previously, when we had early snow, the conditions here were not good for hunting but now it is so mild in September that hunting conditions are very good.

The behaviour of the ptarmigan changes with the snow. Before the snow arrives, the bird hides. When people walk past, it doesn't fly away, it hides. So, you need a dog to locate it, otherwise you can't see it, and you just walk past it. When a dog smells a bird, it stands still, and the hunter sees it. But in winter, the bird is so shy that if you come close, it just flies away. It's much more difficult to hunt in the wintertime, and therefore hunting is more popular before the snow arrives. Nowadays all of September and October are without snow, so a lot of hunters come. They come in the hundreds, and each has one, two or three dogs. They go all over the tundra. But, you know, dog and reindeer is not a good combination. Dogs scare away reindeer.

Reindeer are now accustomed to snow scooters and motorcycles but not to people or dogs. Reindeer are more afraid of people and dogs who come on foot. These dogs are trained not to scare big animals like reindeer. They only go for birds. But, of course, the reindeer doesn't understand that. The dogs run all over the place and make the reindeer nervous. The reindeer are not used to them, so they get scared and run away. This is also becoming a big problem.

Notes

1 Editor's note: Jovsset Ante Sara is a young reindeer herder who was instructed in 2012 by the government of Norway to downsize his herd from 150 reindeer to 75. He did not accept this decision and sued the State. He won his case first at the District Court, and then at the Court of Appeal in Inner Finnmark, who both judged this decision to be a human rights violation. In the process, he gained broad international support as a symbol of Sámi resistance. The State then took the case to the Supreme Court, which decided against Jovsset Ante Sara in December 2017. His case, which is now before the UN Human Rights Commission, illustrates the difficult situation of herders in a democratic country that relies on gas, oil and mineral extraction from lands occupied for millennia by the Indigenous Sámi.

2 As of 17 April 2020, this amounts to approximately 35,435 euros or US$38,500.

3 In the Sámi language, a *siida* (pl. *siidat*) is the traditional cooperative social unit made up of a group of herder families within the reindeer herding community.

11

Herders and Drought in the Sahel of Burkina Faso: Traditional Knowledge and Resilience

HANAFI AMIROU DICKO

My name is Hanafi Amirou Dicko. I am from Burkina Faso, the region of the Sahel most affected by drought. I am the President of Dawla Sahel (Association of the Traditional Herders of Sahel), a member of the Federation of Herders of Burkina and a herder myself (Fig. 11.1). I come from the Sahel; I have never lived anywhere else.

When I was young, I knew the green Sahel, the land of herding *par excellence*. Before 1966, there were no droughts. Its signs were first felt in the years 1965,

Figure 11.1 Hanafi Dicko at the International Conference *Resilience in a Time of Uncertainty: Indigenous Peoples and Climate Change* held at UNESCO during UNFCCC COP 21 (2015).

1966 and 1967. The first time, we told ourselves that it was temporary. But it came back, each time more severe than the last, and now the tragedy has settled in. The great droughts have even forced some herders to become nomads.

How the Drought Arrived

A drought is not something that takes you by surprise. It sets in little by little and announces its arrival. It arrives in stages and each stage is more severe than the previous one. The main sign is the southern wind. In the months of April, May, June and July, when you see the wind blowing from the South to the North, it is not a good sign. A drought is brewing. When we see this, we know there will be a change.

The first stage of the drought arrived in 1968–1969. In the beginning, some people saw a change, but they didn't believe it would last. We were surprised to see the sun unleash all of its fury on Burkina Faso's nature: to see the trees lose their worth: the broadleaf trees first, then the big trees followed. And now, the rich pasturelands are also disappearing. We think it is hot now, but it was much hotter then! It was so hot that when a lizard left one tree to climb another, it could not get there owing to the heat. We found birds that fell from the sky.

That year also marked the beginning of the long transhumance. This transhumance brought some real challenges. Whereas in the past, we were used to vast grazing lands and also pastures reserved for our herds alone, we now had to move South where, as herders from another region, we were looked upon as a source of problems. Some of us even became nomads. A nomad is someone who leaves home, the place of all his memories – of parents, grandparents, children, marriages, etc. – for another destination, fully aware that he may never return.

The second stage of the drought, in 1973–1974, was even more severe than the first. Some far-sighted herders, remembering the drought of 1968–1969, fled early on to the far reaches of southern Burkina Faso. But those who stayed behind were unable to move because their animals were too weak to undertake the transhumance. Starting in April to May, catastrophe hit; our animals died. At that time, when we were already in great despair from the lack of water and pasture for our animals, we were astounded to see huge herds arriving from Mali and Niger: animals by the thousands. Following an old saying of ours that 'herding has no boundaries', we let them stay with us and shared the few resources that were available. However, owing to the scarcity of resources, we all lost our herds. They were dying by the hundreds. All those who could not move south in time lost their animals. Some herders lost everything. Some could not bear it and ended their lives; others lost their sanity. Some left to go live in the city, but without any technical training, they couldn't do anything there. Their training was to follow animals, but there were no more animals. There was nothing but desolation.

The government of the time sent out an international appeal, which provided us with some encouragement. The UN Secretary-General – who back then was Kurt Waldheim – came all the way to Dori in the Sahel of Burkina Faso to assess the tragedy. As far as the eye could see, there were animals lying dead. Later, Mr Jacques Foccart, French Secretary-General for African and Malagasy Affairs from 1960 to 1974, also came to see us. Thereafter, we received international aid in the form of food, such as millet, sorghum and rice, but this was for our consumption only and not for our animals. Our animals are our life, but we were unable to help them.

The third stage of the drought, in 1983–1984, was even more severe because in addition to the previous grim conditions, there was also the wind: the desert storms. When this wind blows, it is first white and close to the ground. If it rises, it becomes red. When it covers nature, everything becomes black because it is full of particles that shroud the sun and create darkness. When it becomes black like that, you can't see more than a metre ahead of you even in a vehicle with fog lights. The wind is stifling. It is hard to breathe. That's the moment when all you can hear are the cries of children and of animals who are out in the bush and want to find their way back home. This wind is preceded by storm clouds that raise your hopes that rain is on the way but then all of a sudden, you notice the whiteness of the wind close to the ground.

Before the first drought, this type of wind was rare. It was in 1983–1984 that it became more and more frequent. It came several times, devastating the vegetation, ripping out the parched shrubs that had been desiccated by the two previous droughts, choking small ponds and filling the large ones with sand. The desolation was total.

We Don't Want to Practise Transhumance

We are a sedentary people. When I was young, before the drought, there were all kinds of trees and grasses in my region. The grasses of the Sahel and the leaves of the trees of the Sahel are richer than those in the southern part of the country. We had no problem with water. The rains were regular throughout the winter season. Back then, the animals didn't need to migrate. They would stay in one place or move some 15–20 km from one village to the next. Before the drought, transhumance was limited. We didn't leave the country or the Sahel region. We, the people of the Sahel, didn't want this transhumance (Fig. 11.2).

At home in the Sahel, there are fewer diseases. For several years in a row, we would have no diseased animals. But down south, there is disease. When you

Figure 11.2 Hanafi Dicko with his herd. [A black and white version of this figure will appear in some formats. For the colour version, please refer to the plate section.]
Photo courtesy of H. Dicko

go to the South, the animals must be vaccinated every time, mostly because of the tsetse flies. We also have to check them to remove ticks. This is not required in the Sahel.

Traditional herding has its fair share of challenges. Everywhere they go, traditional herders are considered strangers, even in their own country. There are always conflicts between farmers and herders, often violent. Back at home, there is grass and arboreal fodder (leaves on trees). We don't want to leave the Sahel.

Climate Change Pushed Us Out of the Sahel

The drought started in the Burkinabe Sahel and at the same time in Mali and Niger. Now we do two types of transhumance: early and seasonal. The herds walk 200–300 km, if not more. They leave from here and can reach Ghana or Côte d'Ivoire. Climate change has pushed us out of the Sahel. You may have a large herd but there is no grass; there are no trees or pasture of any kind. So, you are forced to go looking for grass wherever it happens to be.

We've also witnessed the need for deeper and deeper wells. Previously, the rainy season lasted five months. Now it only lasts two months and during that time, there are no more than six substantial rainfalls for the entire year. So, water infiltration is weak and every year it only gets worse. Earlier, if we dug a well three metres deep, it was sufficient to provide us with water all year round. If you dig a similar well today, after one month you will need to dig deeper and again deeper after two months. In fact, nowadays you need machines to drill wells of wide diameter with cement walls.

We Continue Herding, We Have to Adapt

Nowadays, the signs of drought are permanent. There are no more large trees. There is no grass like there was before. The drought is here to stay. The question now is: how do we adapt?

We have no choice; we have to adapt. We continue herding. Herding is our culture. We are proud to perpetuate herding from one generation to the next. Children are immersed in herding from an early age. They become herders thanks to the knowledge they acquire over time.

But the herders of the Sahel received no international aid to rebuild their herds. With the local knowledge and know-how that herders master, people started to bring back the animals. We don't go and buy animals from the South; it all happens within the Sahel. We sell animals to each other, as other breeds of zebu can't be kept here. The animal breeds we have here are indigenous. We started rebuilding our herds with what animals we had left, and today, *Alhamdulillah*, some have succeeded, albeit not to the same extent as before.

The technique of rebuilding herds is in our head, in our mind, in our body. The people of the Sahel are herders. It is our only activity, and we master it. We possess knowledge about herding and breeding. We know which cow can reproduce, which bull is best prepared for mating, to produce good progeny. In the Sahel, some people are so attached to their herd and know their animals so well, that they don't need anything else to guide them. When a cow is pregnant, a person from the Sahel looks at her udder after it has started to drop. If they tell you, 'This cow, if she gives birth, it will be a bull', then sure enough, the calf turns out to be a bull! And when a bull has just mated with a cow, they watch him come down and they tell you, 'If this cow gives birth, it will be a heifer.' They know it already, from the moment of mating. And they are right. It is their job. They were born into it.

Herders Have also Started Cultivating

Herding has evolved somewhat. Earlier, herders had large herds and farmers had no animals, but now some farmers also have a herd. This is a form of adaptation because not only does it enable them to enrich their fields with organic manure, but it also constitutes a sort of savings that can be mobilized in the event of a bad harvest. Farmers, who were able to buy and maintain animals, became agro-pastoralists. It was an opportunity for them to buy two or three and start herding. Since this land is a herding land *par excellence*, it is easy to enlarge one's herd.

Often, even when they have animals, farmers don't have time to watch them themselves and they entrust them to herders. Farmers and herders are people who live together. Farmers entrust us with animals and help us when we need cash or when something else arises. There is no fixed cost, except when farmers ask us to move our herd to their fields to fertilize them. Then we talk. The pens are made of thorns and moved to another location after a week so that everyone can have manure in their field. Farmers pay a certain amount for each pen on their fields. And if the herders help build the pens, then the farmers must also feed them morning, noon and night.

In the same way, herders also now have small fields. Any person born in a village owns a field, whether they cultivate it or not. Even if they leave for a hundred years, the land remains theirs. Someone can farm it with the herder's agreement, but land can't be acquired just like that. People inherit it from their parents, their grandparents. Just because someone leaves, it does not mean that they will lose their land. In the old days, when there was no drought, they had a field, but they left it for the farmers to cultivate. Now, with the changes, herders have also started cultivating. They're not like real farmers. They cultivate only millet for themselves and give the harvest residues to their animals.

With so little rain, the grass is not able to fully mature. So there is less grass and some varieties have disappeared altogether. Some very rich types, such as *selbere*, *nyomre*, *wadagoore* and *dajje*, have disappeared, or are getting ever scarcer. We are forced to complement the animals' diet. Not the whole herd at once, but each animal at the appropriate time: the pregnant ones, the heifers you are preparing for sale or the bulls who will sire. Their diet is supplemented with cattle cakes. At harvest time, we have to mow the grass, cut millet stalks in the farmers' fields, and store them for the dairy cows and those whose diet we want to reinforce.

As people buy animals for their consumption, a herder can sell an animal to obtain cash to buy cattle cakes. For instance, if after wintering, we want to stock up, we choose a bull or a heifer, sell it, and if feed isn't too expensive at that time, we buy a large quantity and store it. Then in the months of April to May, when times are hard, we start feeding our animals with the stored food. By selling a bull to buy feed, we are able to maintain the herd and manage its diet.

In the Bush, Children Have No Time to Go to School

Our children were born herders. Now it is for them to choose whether they can or want to be herders. They know the life of the animals. If they find something better than herding, they leave. But the one who did not go to school is obliged to follow the animals. Those of us who live around the cities manage to send our children to school. All my children got an education. It is important for us. But it is not the case for all herders. In the bush, children don't have time to go to school since they follow the transhumance every year. Today, this is a major dilemma for herders.

We Will Not Abandon Herding

My son Amadou says, 'If one thing is certain, it is that herding is an ancestral activity that we will not abandon even if we are intellectuals.' There are plenty of educated people back home who have animals. What we say back home is that you always have to have animals. For the wealthiest among us in the Sahel, their fortune started with herding. And those who are rich today have animals: a hundred head. For sure, they knew how to semi-intensify the practice of herding. This is crucial because there is not much grass in the bush. As the animals keep moving around, it wears them out and causes diseases as well. In order to semi-intensify, we bring feed for the animals on site. We sell milk, animals and even manure. So, the concept of business is developing more and more. In the future, no doubt we will progress towards a type of herding that is not entirely traditional, but somewhat modernized.

Taking Things into our Own Hands

I am the president of the Association of the Traditional Herders of Sahel. An association reinforces one's credibility to some extent. After this series of droughts, people have formed clusters and associations to share ideas and activities, and to strengthen their voices. It is better than working on your own. The associations are headed by people living in cities, while in the bush, the herders have organized themselves into clusters. The wives of transhumant herders have also organized clusters that seek support for their ovine and bovine feeding activities. With an association, you may be fortunate enough to reach out to donors who wouldn't trust you as just an individual. Whereas herder clusters are limited in their outreach, people in the associations have been trained. They have opportunities to contact donors. They can bring their issues to the cities. That is the advantage of having an association.

In the times ahead, the most important thing for us is to take things into our own hands. If we count on someone coming to help us, we won't be able to move forward. That is why herders manage on their own, to protect their herds, to conserve them and manage them little by little.

We have the immense hope that one day the cause of traditional herders of the world will be recognized. But we must still be patient. It will still be a while before people can benefit from the COP 21 held in Paris and COP 22 in Marrakech, before their impact will extend to the far corners of Africa. We herders say among ourselves that each of us must manage to maintain our own animals. But if there is help, then all the better.

At the COP meetings I attended, I was really moved to see all these people mobilized around our cause, and personally, I will not forget this. It reassures me and allows me to move forward, facing the hardship that is brought by climate change. With all these people gathered, we know that wherever the problem may be, they will identify and address it. All we need to do is pool our knowledge together and, in this way, identify the best way forward.

When You Are a Herder, It Is Forever

The challenges facing herders around the world are diverse and complex. Whereas some cry out because of rising flood waters, others lament the drought. We are aware that these phenomena are difficult to apprehend for the scientific world and decision-makers. But what we request is assistance and enduring support for herder interests to be taken into account in pasture management policies and most of all, to recognize the value of our local knowledge. That will enable us to carry on traditional herding. We have never given up. When you are a herder, it is forever.

Acknowledgements

Special thanks to Mr Amadou Dicko, Researcher in Animal Production, National Centre for Scientific and Technological Research (CNRST), Institute for the Environment and Agricultural Research (INERA), Burkina Faso.

Part III

Global Change and Indigenous Responses

12

Competing Paradigms of Himalayan Climate Change and Adaptations: Indigenous Knowledge versus Economics

JAN SALICK

Introduction

The eastern Himalaya are critical for understanding the consequences of climate change for bio- and cultural diversity. They are the most linguistically diverse area outside of New Guinea, the most biologically diverse temperate region (Mittermeier et al., 2004; Mutke and Barthlott, 2005) and the most impacted by climate change outside the polar regions (Williams et al., 2007). To date, Himalayan temperatures have increased much faster than the global mean and are predicted to continue increasing at a rate 1.5–1.75 times faster than the global mean in this century (Shrestha et al., 2012; IPCC, 2013). Himalayan precipitation is predicted to increase and to become more variable and extreme (Kohler and Maselli, 2009; Mittal et al., 2014) in an area already famous for torrential monsoons. As ice and snow melt and rains intensify, this 'global water tower' (Immerzeel et al., 2010) threatens to become ever more unstable. Downstream, over a billion people are dependent on the great rivers that originate in the Himalaya and are thereby also made vulnerable to climate change (Füssel, 2007; Xu et al., 2009).

Research on climate change in the Himalaya has centred on glacial retreat (Scherler et al., 2011), the water tower (Xu et al., 2008, 2009; Immerzeel et al., 2010), tree line advance (Wang et al., 2006; Baker and Moseley, 2007; Schickhoff et al., 2015), phenological changes (Yu et al., 2010; Shrestha et al., 2012; Hart et al., 2014), flooding caused by glacial lake outburst (Sharma et al., 2010), local perceptions (Ives, 2004) and paleoenvironment (Herzschuh et al., 2010; Kramer et al., 2010). Recently, we[1] have been providing data on the effects of climate change on the exceptional alpine biodiversity of the Himalaya (Salick et al., 2014, 2019) and on Himalayan peoples including Tibetans, a culture already under threat (Salick et al., 2005, 2007, 2013, 2014, 2017; Salick and Byg, 2007; Byg and Salick, 2009; Salick and Ross, 2009a, 2009b; Byg et al., 2010; Salick, 2012; Salick and Moseley, 2012; Konchar et al., 2015). Here, I review this research and reconsider it in light of two competing visions: Indigenous traditions and market economics.

Methods

To understand how climate change threatens the Himalaya and their peoples, we have carried out long-term research on the ethnobotany of Himalayan climate change, integrating both long-term alpine vegetation studies and participatory research in Tibet, Bhutan, Nepal and, most recently, Pakistan. In the year 2000, we began ethnobotany research in the Tibetan Autonomous Prefecture, Yunnan, China (TAP, along with the Tibetan Autonomous Region, henceforth referred to simply as 'Tibet') using diverse participatory methodologies including mapping, calendars, photography and interviews. In 2004, we joined the Global Observational Research Initiative in Alpine Environments (GLORIA: www.gloria.ac.at) and began intensive alpine vegetation data collection in the Himalaya to monitor the effects of climate change over time (Fig. 12.1; Salick et al., 2009, 2014, 2019; Pauli et al., 2015). These methods have been reproduced along a 1500 km transect from the eastern borderlands of Tibet through Bhutan and Nepal into the Central Himalaya and expanded more recently into the Western Himalaya in Pakistan. Long-term research continues and the results of these research studies have been and continue to be published in the climate change literature.

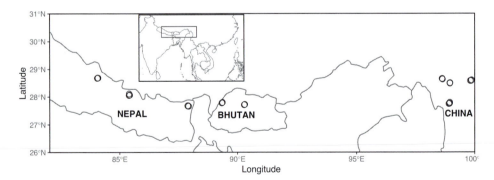

Figure 12.1 1,500 km transect with 9 major target regions across the eastern Himalaya. Each target region includes 3–4 mountain summits 3,800–5,000 masl and covers a range of ecotones: subalpine-lower alpine; lower alpine-upper alpine; upper alpine-subnival; and subnival-nival.

Results

Briefly summarized, Himalayan alpine vegetation is responding to climate change with increasing plant abundance, richness and diversity (Salick et al., 2014, 2019). Increases are especially pronounced for Himalayan endemic plants, a result not seen in other GLORIA regions. However, simultaneously with these increases, Himalayan 'sky islands' – isolated elevational vegetation zones on mountain peaks – are shrinking in size as temperatures warm and alpine plant distributions move upwards (Fig. 12.2). As a result, the long-term future of Tibetan alpine vegetation is threatened by climate change, even as the short-term processes remain somewhat contradictory.

Participatory research with Tibetans and other Himalayan peoples in Bhutan, Nepal and Pakistan has revealed their keen perceptions of climate change impacts with nuanced adaptations and mitigations (Salick et al., 2005, 2014, 2017; Byg and Salick, 2009; Salick and Moseley, 2012; Konchar et al., 2015; Salick, 2015; Hart and Salick, 2017). Himalayan peoples recount rapid changes in Himalayan

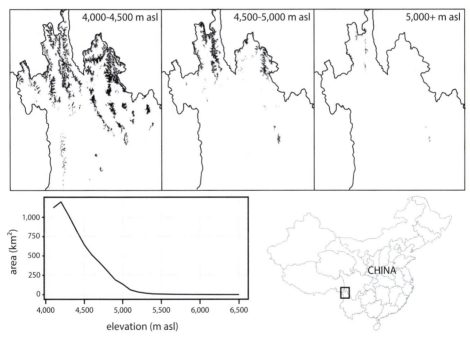

Figure 12.,2 Shrinking 'Sky Islands' of southeastern Tibet: eastern Himalaya of the Tibetan Autonomous Prefecture, Hengduan Mountains, Yunnan, China. With climate warming, alpine habitats (dark areas on the maps) move up mountains and their area decreases precipitously. A 500 m upward habitat shift (the difference between each map panel) would take place in 55–100 years at the present rate of warming: and that rate is predicted to increase.

environments with major consequences for traditional agropastoralism, as well as cultural and spiritual adaptations. These, in turn, reveal dramatic tipping points – when small or gradual changes build up to cause catastrophic change – of production and land-use systems, of subsistence and cash or tourist economies, and of traditional and contemporary cultural orientations. Before comparing paradigms with which to interpret the responses of Himalayan peoples to climate change, I will first review results on land use and cultural and spiritual adaptions in greater detail.

With climate change – especially warming temperatures – land use, both agriculture and extraction of non-timber products, is shifting towards cash endeavours. The most frequently reported effects are agricultural challenges and changes. Crops are changing rapidly, and the number of crops and new varieties are increasing, making Himalayan agricultural systems more complex for traditional farmers with much to learn and to innovate. New fruit and vegetable crops are being grown and commodities produced. However, these are mostly for tourists or for sale, as locals generally retain their traditional diets. These new, introduced varieties are replacing traditional Himalayan varieties developed locally. Crops such as rice, maize, vegetables, walnuts and fruits are being grown at higher elevations. Some traditional crops are being abandoned, including culturally important and nutritious buckwheat, a high elevation and very labour intensive crop, which is now being replaced by either lower elevation produce with expanding ranges or by other activities such as tourist trade, gathered products and cash labour.

Traditional yak herding, which is time consuming and is done at ever higher elevations as a result of climate change, is now being replaced by lucrative gathering, cash agriculture and tourism. Harvesting of non-timber products, including the profitable *chung cao* (*Ophiocordyceps sinensis*) which was once traditional but is now commercial, dominates the early spring calendar (Salick et al., 2005, Konchar et al., 2015). Also gaining in importance and time allocation later in the year is the collection of medicinal plants such as Snow Lotus (Law and Salick, 2005, 2006) and mushrooms like the *matsutake* (*Tricholoma matsutake*; Salick et al., 2005; Amend et al., 2010). In general, seasonal activities are changing with climate. The traditional Tibetan calendars, produced by specialized Tibetan astrologers, are becoming difficult to use and the astrologers struggle to adjust (Salick et al., 2017). Elders, who have traditional knowledge about and experience with weather and planting, can no longer predict when to plant.

Indicator species are used to decide when to conduct agricultural activities (Salick et al., 2005). Himalayan Indigenous peoples report that indicators are failing with climate change. Indicator flowers, such as rhododendrons, bloom earlier (Hart et al., 2014; Hart and Salick, 2017, 2018) and are not in synchrony with agricultural activities. Indicator animals, such as cranes, stay longer and eat grain; cuckoos are heard at unusual times of year (Salick et al., 2017). New cash crops have yet to be, if ever, associated with the natural Himalayan world.

Forests are expanding to higher elevations. Warming and both Chinese and traditional Tibetan bans on tree cutting have been resulting in increasing forest cover and carbon sequestration (Salick and Moseley, 2012). Traditional conservation strategies such as sacred sites and sacred mountains (Anderson et al., 2005; Salick et al., 2007; Salick and Moseley, 2012; Salick et al.,2019) also effect carbon sequestration, as does agroforestry with traditional crops such as walnuts, and new fruit crops that thrive in the warming climate.

Cultural practices are changing with climate (Byg and Salick, 2009; Konchar et al., 2015; Salick et al., 2017). With increased precipitation, traditional flat mud roofs are collapsing and are being replaced with purchased corrugated roofing, pitched to shed the increasing rainfall. With climate change, traditional yak robes which were fashioned by hand have become too warm, while contemporary, light weight hiking gear is nationally produced and is inexpensive but must be purchased. Road incursions into the Himalaya bring more tourists, tourism conglomerates and commercial goods, while undermining traditional ways of life and increasing carbon footprints. Thus, adaptation to climate change, loss of traditional cultural practices and growth of cash economies all become inextricably intertwined.

The health of plants, animals and people is affected by climate change according to Himalayan Indigenous peoples (Salick et al., 2005, 2017). There are increasingly serious pest attacks attributed to changing weather that favours diseases and insects. These are also connected with Tibetan beliefs that unclean and incorrect living (spiritual pollution) is punished by plagues. For crops, generally, insect attacks are frequent when it is unusually hot and dry, while diseases flourish when rainy conditions persist. New weeds that villagers have never seen before are appearing and new insects are causing problems, especially with new crops; for example, wasps are becoming pests of the expanding grape crop. Mice are no longer killed by severe winters and are multiplying out of control. Farm animals, particularly pigs, are succumbing to new and more virulent diseases. Flies and mosquitoes are also multiplying quickly and moving to ever higher elevations. Malaria is reportedly moving into lower Himalayan foothills. Refrigeration appliances are increasingly purchased, both for tourists and to avoid food spoilage that is a new problem with warming and non-traditional diets.

Spiritual responses to climate change are evidenced among Indigenous cultures around the world and certainly no less in Tibet and the Himalaya, where religion is paramount (Anderson et al., 2005; Salick et al., 2005, 2007, 2013, 2019; Byg and Salick, 2009; Salick and Moseley, 2012). People are both relying more heavily on prayer and supplication to evade climate catastrophes and are worried about being forsaken by their gods as the snows melt on sacred mountains and as sacred crane populations diminish. Simultaneously, young people heavily involved in cash economies may no longer see the value of traditional religion.

Adaptation strategies in the high Himalaya include a shift from traditional agropastoral practices to a more diversified blend of agropastoralism, tourism services and cash-crop production (Salick et al., 2005, 2017; Konchar et al., 2015). Climate change has tipped the scales in favour of the production of fruits and vegetables, cash crops previously unsuited to the local climate. Diversification of livelihood strategies signifies transformation within the socio-ecological system of the high Himalaya. This may enable greater resilience to long-term climatic change, but may also move Himalayan peoples away from Indigenous knowledge and distinctive cultural traditions. Analyses of the effects of and adaptations to climate change in the Himalaya differ dramatically depending on whether analysts use an economic paradigm or view it from the perspective of Indigenous knowledge.

A Case Study of Tibetan Wine

An outstandingly successful adaptation to climate change near the sacred Mt Khawa Karpo is grape and wine production. In general, it illustrates the competing visions of climate change and their implications for local adaptation. As local Tibetans recount, in the mid-nineteenth century, French missionaries established a lovely Catholic church in Cizhong, just south of Mt Khawa Karpo (Fig. 12.3a). Since 2006, the church is listed as a national historic and cultural site. Its ceiling vault was beautifully illuminated by the monks, who were also avid botanists, with colourful Himalayan wildflowers (Fig. 12.3b). These monks and visiting French botanists collected some of the earliest botanical specimens from the area (available at the National Museum of Natural History (MNHN) in Paris and elsewhere in France) from which many new genera and species were named. One example is the famous genus of Blue Poppies, *Meconopsis* Vig., described in 1814 by the French botanist Louis Guillaume Alexandre Viguier in Montpellier. Within the high walled churchyard, the French priests and monks grew Cabernet Sauvignon grapes that they had brought from France for communion as well as for their personal consumption. The missionaries taught local villagers to produce wine in the French tradition, making Cizhong the only Himalayan area with a historical wine culture. In 1952, the Cizhong vicar Francis Goré finally abandoned the church and fled to Hong Kong as the Communists formed the People's Republic and exerted control over the borderlands (Salick and Moseley, 2012). Nonetheless, the local Tibetans, trained by the French, continued to manage the

Figure 12.3 Case study of Chizong wine production, Tibetan Autonomous Prefecture, Yunnan, China: (a) French missionaries established a lovely Catholic church in Chizong on the southeastern edge of the Tibetan plateau; (b) the ceiling vault depicts endemic alpine flowers of the eastern Himalaya first described by French botanists; and (c) Tibetan ice wine production won international acclaim. [A black and white version of this figure will appear in some formats. For the colour version, please refer to the plate section.]

small vineyard within the walled, warm microclimate. As climate change set in, the warmth outside the church walls became suitable for grape production at a larger scale. Tibetans distributed cuttings of the grapes, and vineyards thrived throughout the area. Grapes became a very successful cash crop, completely displacing traditional Tibetan agriculture in some particularly suitable villages. But Tibetans were not satisfied with this success; they had seen wine produced and consumed and they were determined to add value. Tibetan wineries were established and Tibetans went to Europe to learn more. In European mountain areas where frosts and snow are rapid, they saw ice wines being produced. Ice wine is a sweet dessert wine produced from grapes that have been frozen while still on the vine. Ice wine takes delicate management and the grapes, once frozen, must be harvested and processed immediately. The Tibetans near Mt Khawa Karpo developed such excellent ice wine that they won second prize in an international ice wine contest with their 'Sun Spirit' label (Fig. 12.3c).

However, this local success story also has a downside. The prosperity of wine making in the area attracted large-scale investors and the Chinese government, who have since taken over most of the added value production at an industrial scale, leaving villagers to grow the grapes only, with little control over production, product or price. Villagers' profits are greatly reduced as a result and their workload considerably increased.

Discussion

Our studies and the results summarized in the previous section have been interpreted differently depending on points of view. Economic and Indigenous knowledge perspectives seem to be particularly disparate. Among a group of conservative businessmen, I was told, 'Any natural event has some winners and some losers. Himalayan peoples seem to be winning.' (St. Louis Rotary 2016, pers. comm.) They were referring to increasing biodiversity, agricultural capacity and options to enter cash economies, even though, at the same time, they denied human-induced climate change. These benefits are real. As described, unlike in other parts of the world, increasing frequencies of plants, numbers of plant species, plant diversity and frequencies of endemic plants have been reported in alpine habitats (Salick et al., 2019). Crops that are being grown in the Himalaya are increasing dramatically and rapidly, to the point that we propose a climate-driven tipping point (Konchar et al., 2015). Fruits and vegetables are providing substantial cash incomes in some areas of the Himalaya, including our case study of Tibetan grape and wine production in the eastern borderlands of the Tibetan plateau. Cash incomes also drive extractive harvests of non-timber products including *chung cao* and *matsutake* mushrooms, both known as aphrodisiacs and sold to wealthy Chinese and other Asian men. Reciprocally, the sacred Snow Lotus is sold to women for 'women's problems'. But do these economic benefits also come at a cost?

The economists' paradigm and analyses unsettled me because of my own bias towards Indigenous knowledge and culture and against economic disenfranchisement. I see ancient traditions being lost and made irrelevant. I see Himalayan peoples entering the global economy on the bottom rung with little opportunity for advancement. One Tibetan remarked, 'Climate change may annihilate Tibetan culture once and for all.' (Lhasa 2012, pers. comm.) Loss of traditions – from crops, food, annual cycles, material culture, traditional practices and techniques to spiritual beliefs – are indirect but profound effects of climate change. We see the dominance of a cash and global economy moving in with climate change. When Himalayan peoples grow for sale the fruits and vegetables newly available to them, there is less time and labour available for traditional food production. Yak grazing, an iconic high Himalayan land use, is now taking place on increasingly higher elevation alpine pastures and on diminishing land areas owing to climate change. There is a subsequent decline in the production of yak products: milk, cheese, meat, woven/knitted/felted yak hair and yak leather products – fundamentals of traditional high Himalayan society. Genetic diversity and production of traditional crops such as buckwheat, barley and walnuts are declining as farmers switch to cash crops, thereby threatening nutrition and traditions.

From an economic point of view, the conversion of Tibetan villages to vineyards represents a substantial economic opportunity made possible by climate change. However, I also see the disappearance of traditional agriculture, increases of store-bought foods lacking in nutrition and cheap polyester clothing, along with concomitant waning of Tibetan Buddhism and traditional spiritualism. Economy of scale is the economists' explanation for government and conglomerate takeover of wine production, while I see institutionalized racial and economic inequality obstructing Indigenous rights and justice.

These competing paradigms are omnipresent in global climate change policy as well. Rio+20 or the 2012 United Nations Conference on Sustainable Development in Brazil took on the theme of Green Economics (transliterated to 'Greed Economics' by protestors), whereas the original Rio conference in 1992 highlighted Indigenous peoples' rights – a dramatic paradigm shift brought about by a power elite. Following Rio+20, the 2030 Agenda for Sustainable Development was drafted, again with little attention to Indigenous peoples and with a significant focus on economics. Indigenous peoples are recognized only as vulnerable and lumped with children and the elderly, the disabled, HIV/AIDS infected and refugees, rather than recognized as the creative, innovative force for change that we know they are. Throughout policy formation on climate change, little attention has been paid to the plight, rights or justice due to Indigenous peoples – let alone to their positive contributions (Salick and Byg, 2007; Salick and Ross, 2009a, 2009b). In contrast, there is major attention focused on the economic impacts of climate change. To circumvent the economic emphasis of both sustainability and climate change policy, many Indigenous and traditional groups concentrate on 'sovereignty'. Sovereignty applies the rights of peoples, communities and countries to define their own healthy and culturally appropriate goals and ecologically sound policies, rather than presuming economics as either the metric or the goal. Such a paradigm shift can only take place with the participation, decision-making and leadership of well-informed local peoples.

Continued development of relevant, place-based adaptations to, and mitigations of, rapid Himalayan climate change depends on local peoples' ability to understand the potential impacts of climate change and adjust within complex, traditional socio-ecological systems and to understand the impacts of a global economy on their lives and cultures (Fenstad et al., 2002a, 2002b). Himalayan communities must be given opportunities to voice their observations of climate change, localized concerns, and culturally relevant adaptations and mitigations in global discussions of development and climate change policy. Human rights and justice are in the balance and are the only option for cultural continuance and sovereignty in the face of unprecedented climate change and global economics.

Note

1 During more than fifteen years of field research on climate change and adaptations in the Himalaya, teams of people have worked on my projects, including from the Missouri Botanical Garden, St. Louis, now under the leadership of Dr Robbie Hart; the Shangri-la Alpine Botanical Garden, Tibetan Autonomous Prefecture, Yunnan, China, under the direction of Mr Fang Zhendong; National Biodiversity Centre, Bhutan, under the leadership of Ms Sangay Dema; and Department of Botany, Tribhuvan University, Nepal, under the leadership of Dr Suresh Ghimire (see previous publications). My intense gratitude and devotion go to these inspired people and quality institutions.

References

Amend, A., Fang, Z., Yi, C. and McClatchey, W. C. 2010. Local perceptions of Matsutake mushroom management, in NW Yunnan, China. *Biological Conservation*, 143: 165–172.

Anderson, D., Salick, J., Moseley, R. K. and Ou, X. 2005. Conserving the sacred medicine mountains: A vegetation analysis of Tibetan sacred sites in Northwest Yunnan. *Biodiversity and Conservation*, 14: 3065–3091.

Baker, B. B. and Moseley, R. K. 2007. Advancing treeline and retreating glaciers: Implications for conservation in Yunnan, P.R. China. *Arctic, Antarctic, and Alpine Research*, 39: 200–209.

Byg, A. and Salick, J. 2009. Local perspectives on a global phenomenon: Climate change in Eastern Tibetan villages. *Global Environmental Change*, 19: 156–166.

Byg, A., Salick, J. and Law, W. 2010. Medicinal plant knowledge among lay people in five eastern Tibet villages. *Human Ecology*, 38: 177–191.

Fenstad, J. E., Hoyningen-Huene, P., Hu, Q., Kokwaro, J., Nakashima, D., Salick, J., Shrum, W. and Subbarayappa, B. V. 2002a. Science, traditional knowledge and sustainable development. *ICSU Series on Sustainable Development*, No. 4, Paris.

Fenstad, J. E., Hoyningen-Huene, P., Hu, Q., Kokwaro, J., Salick, J., Shrum, W., Subbarayappa, B. V. and Nakashima, D. 2002b. *Science and Traditional Knowledge*. Paris: ICSU. www.icsu.org/Gestion/img/ICSU_DOC_DOWNLOAD/ 220_DD_FILE_Traitional_Knowledge_report.pdf

Füssel, H-M. 2007. Vulnerability: A generally applicable conceptual framework for climate change research. *Global Environmental Change*, 17: 155–167.

Hart, R. and Salick, J. 2017. Dynamic ecological knowledge systems amid changing place and climate: Mt. Yulong rhododendrons. *Journal of Ethnobiology*, 37: 21–36.

Hart, R. and Salick, J. 2018. Vulnerability of phenological progressions over season and elevation to climate change: Rhododendrons of Mt. Yulong. Perspectives in Plant Ecology. *Evolution and Systematics*, 34: 129–139.

Hart, R., Salick, J., Ranjitkar, S. and Xu, J. 2014. Herbarium specimens show contrasting phenological responses to Himalayan climate. *PNAS*, 111: 10615–10619.

Herzschuh, U., Birks, H. J. B., Ni, J., Zhao, Y., Liu, H. and Liu, X. 2010. Holocene land-cover changes on the Tibetan Plateau. *The Holocene*, 20: 91–104.

Immerzeel, W., van Beek, L. and Bierkens, M. 2010. Climate change will affect the Asian water towers. *Science*, 328: 1382–1387.

IPCC. 2013. *Climate Change: The Physical Science Basis*. Cambridge: Cambridge University Press.

Ives, J. 2004. *Himalayan Perceptions*. London: Routledge.

Kohler, T. and Maselli, D. 2009. *Mountains and Climate Change: From Understanding to Action*. Bern, Switzerland: Geographica Bernensia.

Konchar, K., Staver, B., Salick, J., Chapagain, A., Joshi, L., Karki, S., Lo, S., Paudel, A., Subedi, P. and Ghimire, S. 2015. Adapting in the shadow of Annapurna: A climate tipping point. *Journal of Ethnobiology*, 35: 449–471.

Kramer, A., Herzschuh, U., Mischke, S. and Zhang, C. 2010. Holocene treeline shifts and monsoon variability in the Hengduan Mountains (southeastern Tibetan Plateau), implications from palynological investigations. *Palaeogeography, Palaeoclimatology, Palaeoecology*, 286: 23–41.

Law, W. and Salick, J. 2005. Human-induced dwarfing of Himalayan Snow Lotus (*Saussurea laniceps* (Asteraceae)). *PNAS*, 102: 10218–10220.

Law, W. and Salick, J. 2006. Comparing conservation priorities for useful plants among botanists and Tibetan doctors. *Biodiversity and Conservation*, 16: 1747–1759.

Mittal, N., Mishra, A., Singh, R. and Kumar, P. 2014. Assessing future changes in seasonal climatic extremes in the Ganges river basin using an ensemble of regional climate models. *Climatic Change*, 123: 273–286.

Mittermeier, R. A., Robles, Gil P., Hoffmann, M., Pilgrim, J., Brooks, T., Mittermeier, C. G., Lamoreux, J. and da Fonseca, G. A. B. 2004. *Hotspots Revisited: Earth's Biologically Richest and Most Endangered Ecoregions*. Mexico City: CEMEX.

Mutke, J. and Barthlott, W. 2005. Patterns of vascular plant diversity at continental to global scales. *Biologiske Skrifter*, 55: 521–531.

Pauli, H., Gottfried, M., Lamprecht, A., Niessner, S., Rumpf, S., Winkler, M., Steinbauer, K. and Grabherr, G. (eds.) 2015. *The GLORIA Field Manual: Standard Multi-Summit Approach, Supplementary Methods and Extra Approaches*. 5th edition. Vienna: GLORIA-Coordination, Austrian Academy of Sciences & University of Natural Resources and Life Sciences.

Salick, J. 2012. Indigenous peoples conserving, managing and creating biodiversity. In Gepts, P., Famula, T. R., Bettinger, R. L., Brush, S. B., Damania, A. B., McGuire, P. E. and Qualset, C. O. (eds.) *Biodiversity in Agriculture: Domestication, Evolution, and Sustainability*. Cambridge: Cambridge University Pres.

Salick, J. 2013. Indigenous knowledge integrated with long term ecological monitoring: Himalaya. In Thaman, R., Lyver, P., Mpande, R., Perez, E., Cariño, J. and Takeuchi, K. (eds.) *The Contribution of Indigenous and Local Knowledge Systems to IPBES: Building Synergies with Science*. Paris: UNESCO/UNU.

Salick, J. 2015. Ethnobotany integrated with GLORIA. In Pauli, H., Gottfried, M., Lamprecht, A., Niessner, S., Rumpf, S., Winkler, M., Steinbauer, K. and Grabherr, G. (eds.). *The GLORIA Field Manual: Standard Multi-Summit Approach*. 5th edition. Vienna: Austrian Academy of Sciences & University of Natural Resources and Life Sciences.

Salick, J. and Byg, A. 2007. *Indigenous Peoples and Climate Change*. Oxford: Tyndall Centre. http://tinyurl.com/salickbyg2007

Salick, J. and Moseley, R. 2012. Khawa Karpo: Tibetan tradition knowledge and biodiversity conservation. *Monographs in Systematic Botany*, 121: 273.

Salick, J. and Ross, N. (eds.) 2009a. Traditional Peoples and Climate Change. *Special Issue: Global Environmental Change*, 19.

Salick, J. and Ross, N. 2009b. Traditional Peoples and Climate Change. *Traditional Peoples and Climate Change. Special Issue: Global Environmental Change*, 19: 137–139.

Salick, J., Byg, A. and Bauer, K. 2013. Contemporary Tibetan cosmology of climate change. *Journal for the Study of Religion, Nature and Culture*, 6: 552–577.

Salick, J., Byg, A. and Konchar, K. 2017. Innovation and adaptation in Tibetan land use and agriculture coping with climate change. *In Indigenous Peoples, Marginalized Populations and Climate Change*. Cambridge: Cambridge University Press.

Salick, J., Fang, Z. D. and Byg, A. 2009. Tibetan ethnobotany and climate change in the eastern Himalayas. *Global Environmental Change*, 19: 147–155

Salick, J., Fang Z. D. and Hart, R. 2019. Rapid change in eastern Himalayan alpine flora with climate change. *American Journal of Botany*, 106(4): 1–11.

Salick, J., Hart, R. and Li, S. In prep. Naxi Cosmology of Mt. Yulong Sacred Sites, NW Yunnan China and Potential for Conservation.

Salick, J., Yang, Y. P. and Amend, A. 2005. Tibetan land use and change in NW Yunnan. *Economic Botany*, 59: 312–325.

Salick, J., Ghimire, S., Fang Z. D., Dema, S. and Konchar, K. 2014. Himalayan alpine vegetation, climate change and mitigation. *Climate Change Special Volume: Journal of Ethnobiology*, 34: 276–293.

Salick, J., Amend, A., Anderson, D., Hoffmeister, K., Gunn, B. and Fang, Z. D. 2007. Tibetan sacred sites conserve old growth trees in the eastern Himalayas. *Biodiversity and Conservation*, 16: 693–706.

Salick, J., Byg, A., Amend, A., Gunn, B., Law, W. and Schmidt, H. 2006. Tibetan medicine plurality. *Economic Botany*, 60: 227–253.

Scherler, D., Bookhagen, B. and Strecker, M. R. 2011. Spatially variable response of Himalayan glaciers to climate change affected by debris cover. *Nature Geoscience*, 4: 156–159.

Schickhoff, U., Bobrowski, M., Böhner, J., Bürzle, B., Chaudhary, R. P., Gerlitz, L., Heyken, H., Lange, J., Müller, M., Scholten, T. and Schwab, N. 2015. Do Himalayan treelines respond to recent climate change? An evaluation of sensitivity indicators. *Earth System Dynamics*, 6: 245–265.

Sharma, E., Chettri, N., Tse-ring, K., Shrestha, A. B., Fang J., Mool, P. and Eriksson, M. 2010. *Climate Change Impacts and Vulnerability in the Eastern Himalayas*. Nepal: ICIMOD Kathmandu, Nepal. 32pp.

Shrestha, U. B., Gautam, S. and Bawa, K. S. 2012. Widespread climate change in the Himalayas and associated changes in local ecosystems. *PLoS ONE*, 7: e36741.

Wang, T., Zhang, Q. B. and Ma, K. 2006. Treeline dynamics in relation to climatic variability in the central Tianshan Mountains, north-western China. *Global Ecology and Biogeography*, 15: 406–415.

Williams, J. W., Jackson, S. T. and Kutzbach, J. E. 2007. Projected distributions of novel and disappearing climates by 2100 AD. *PNAS*, 104: 5738–5742.

Xu, J., Grumbine, R. E., Shrestha, A., Eriksson, M., Yang X., Wang, Y. and Wilkes, A. 2009. The melting Himalayas: Cascading effects of climate change on water, biodiversity, and livelihoods. *Conservation Biology*, 23: 520–530.

Xu, X., Lu, C., Shi, X. and Gao, S. 2008. World water tower: An atmospheric perspective. *Geophysical Research Letters*, 35: L20815.

Yu H., Luedeling, E. and Xu, J. 2010. Winter and spring warming results in delayed spring phenology on the Tibetan Plateau. *PNAS*, 107: 22156.

13

Coping with a Warming Winter Climate in Arctic Russia: Patterns of Extreme Weather Affecting Nenets Reindeer Nomadism

BRUCE C. FORBES, TIMO KUMPULA, NINA MESCHTYB, ROZA LAPTANDER,
MARC MACIAS-FAURIA, PENTTI ZETTERBERG, MARIANA VERDONEN,
ANNA SKARIN, KWANG-YUL KIM, LINETTE N. BOISVERT,
JULIENNE C. STROEVE AND ANNETT BARTSCH

Introduction

Arctic Indigenous peoples live in close relationship with their environment and experience 'change' over a wide range of spatial and temporal scales (Crate et al., 2010). High latitude warming has significantly exceeded that recorded elsewhere on the planet in recent decades (Stroeve et al., 2012; Larsen and Anisimov, 2014). Hunters and herders from many Arctic regions have reported symptoms of accelerating change, even while characterizing extreme weather events as being, to a great extent, 'normal' in the context of their lives and livelihoods on the land and/ or sea (Krupnik and Jolly, 2002; Forbes and Stammler, 2009; Larsen and Anisimov, 2014). In the Eurasian Arctic, rapid sea ice retreat and thinning in the Barents and Kara Seas have been flagged as critical components of feedbacks to Arctic and global climate change (Stroeve et al., 2012; Kim et al., 2016). During this same period, autumn and winter rain-on-snow events resulting in ice-encrusted, and thus inaccessible, pastures and mass starvation of reindeer (*Rangifer tarandus*) have increased in frequency and intensity across the Low Arctic in northwest Russia (Bartsch et al., 2010; Sokolov et al., 2016). This region is home to the world's largest and most productive reindeer herds. Warmer/wetter winters have negatively affected the much smaller wild reindeer populations on the High Arctic Svalbard (Hansen et al., 2011, 2014); however, these reindeer are neither subject to hunting nor herding. In the Low Arctic, where tundra nomadism remains an important livelihood, there is an urgent need to understand whether and how regional sea ice loss is driving major rain-on-snow events, and the implications of such events for the region's ancient and unique social-ecological systems.

At the circumpolar level, it has been speculated that the greening of the Arctic, resulting in increased tundra vegetation productivity, is tied to sea ice decline and thinning (Bhatt et al., 2010; Post et al., 2013). However, evidence based on tundra shrub annual growth from the Nenets Autonomous Okrug[1] (NAO) and Yamal-

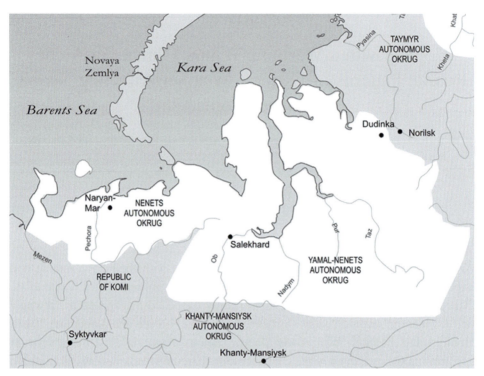

Figure 13.1 Reindeer herding territories (in white) of the east European Arctic and west Siberia.

Nenets Autonomous Okrug (YNAO; Fig. 13.1) does not support such a linkage in the Barents and Kara Seas region. Instead, the trend of increasing deciduous shrub growth appears to be closely tied to the increased frequency and intensity of summer high-pressure systems over West Siberia in recent decades (Macias-Fauria et al., 2012). While attention has been focused on summer warming trends coupled with decreasing access to summer reindeer pastures owing to incremental hydrocarbon development (Forbes et al., 2009; Forbes, 2013), it is winter warming that has actually led to the most significant changes in regional reindeer populations (Moscow Times, 2014; Anonymous, 2015; Sokolov et al., 2016; Staalesen, 2016).

To remain viable, tundra Nenets reindeer nomadism depends on unfettered access year-round to both reindeer pastures and fishing resources (Forbes et al., 2009; Forbes, 2013). Among the approximately 30,000 Indigenous Yamal Nenets of West Siberia, around 6,000 are reindeer nomads who migrate up to 1200 km annually between lichen woodlands (winter pastures) south of the Ob River and northern shrub-graminoid tundra (spring/summer/autumn pastures) on the Yamal Peninsula (Forbes et al., 2009). The total reindeer population in YNAO was

705,000 in January 2016, of which 394,000 were privately owned. The migratory units range in size from 100 to several thousand animals.

In this chapter, we consider how recent and historic rain-on-snow events have shaped Nenets' strategies for coping with extreme weather. In particular, we review evidence for cold season atmospheric warming and precipitation increases over Arctic coastal lands in proximity to Barents and Kara seas and investigate whether these might be related to sea ice loss. Particular attention is paid to a major rain-on-snow event during the autumn/winter of 2013–2014 that led to the starvation of 61,000 reindeer out of a population of approximately 275,000 animals on Yamal Peninsula (Moscow Times, 2014). Rain-on-snow events, when rain on the snow surface refreezes causing a hard crust to be formed, limits access to winter forage beneath and can be devastating. Since records began nearly 100 years ago, this is the region's largest recorded mortality episode. It appears to be part of a pattern of more frequent and intense autumn/winter rain-on-snow events. What is driving this increase is not clear, but the event will be considered here from the perspective of recent literature, including modelling efforts, and direct measures of sea ice and associated atmospheric conditions. If sea ice loss is driving increasingly severe rain-on-snow events and high reindeer mortality, it has serious implications for the future of tundra Nenets nomadism. We also explore Nenets' strategies for adaptation to rain-on-snow events in recent decades and into the future.

Materials and Methods

Between March 2014 and April 2016, we surveyed a total of sixty herders and administrators representing the Yarsalinskii *sovkhoz* (which has the longest migration route) and the Baidaratskaii *sovkhoz* (Fig. 13.2). A *sovkhoz* in Russia is a territory of mixed collective and private herd ownership. Informants included adult men (about 70%) and women (30%), in the age cohorts of 60+ years (40%), 50–60 years (40%) and less than 40 years old (20%). They represented both private and collectively managed reindeer units. We also engaged in intensive participant observation in all seasons with Nenets nomads, and assembled, in this manner, a detailed oral history of herding strategies and movements over several decades. Primary topics for discussion with the Nenets included not only the most recent rain-on-snow event and resulting massive reindeer mortality of the autumn/winter of 2013–2014 but also the pattern of events during the last century and nomadic responses in space and time. Secondary topics included the different interpretations provided by Nenets herders to explain what caused the catastrophic winter of 2013–2014 with its exceptionally high reindeer mortality.

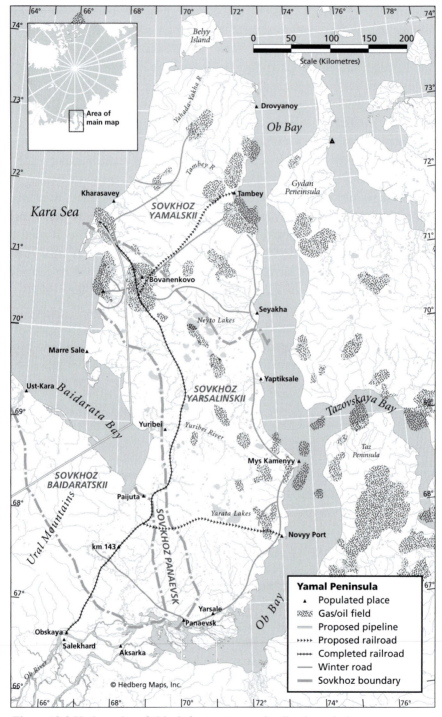

Figure 13.2 Hydrocarbon fields, infrastructure and collective reindeer management (*sovkhoz*) territories on Yamal Peninsula, West Siberia.

With regard to meteorological data, we combed the archives of Atmospheric InfraRed Sounder (AIRS) to search for both seasonal and date-specific anomalies in atmospheric conditions over the Barents and Kara Seas in 2006–2007 and 2013–2014. We checked for precipitation anomalies using ERA-Interim data, a global atmospheric reanalysis produced by the European Centre for Medium-Range Weather Forecasts (ECMWF). The data archives of the Advanced Scatterometer (ASCAT) were similarly searched to see if the autumn 2013 rain-on-snow event could be detected over land.

Discussions with herders and administrators were guided geographically by employing a combination of: (1) large (1:100,000) to medium (1:500,000) scale topographic maps with place names; and (2) moderate (Landsat) to very high-resolution (WorldView, GeoEye, QuickBird2) satellite imagery. The findings from the combined quantitative and qualitative (i.e., participatory) approaches are supplemented with a literature review that focuses on modelling Barents and Kara sea ice loss.

Results

Herders describe a *sered po* (catastrophically bad or distress year in Nenets language) as a year in which a major rain-on-snow event results in snow cover characterized by *gololëd*, a frozen or glazed ice crust on the pasture. According to the herders, the most catastrophic gololëd event of the past century began around 8–9 November 2013 with about 24 hours of rain, after which temperatures dropped and remained below freezing for the remainder of autumn and throughout the winter. ASCAT data accurately detected the severe icing of pastures beginning 10 November 2013 over most of the southern Yamal Peninsula, an area covering about 27,058 km^2 (Fig. 13.3). One herder in his sixties observed that this was 'the worst *gololëd*' in his life and consequently, therefore, since the 1950s. Another herder in the same age cohort explained that,

It happened already in the fall, when there was hardly any snow, and the rain froze right on the pasture. Then on [top of] that it snowed, and then froze. That [surface] is stone hard and impossible to dig through for the animals.

Meteorological data for the above dates revealed the following: Between 5–10 November 2006, the dates of the 2006–2007 winter rain-on-snow event, and from 5–10 November 2013, Special Sensor Microwave Imager/Sounder (SSMIS) shows that Barents and Kara sea ice was steadily decreasing (Fig. 13.4). AIRS data for autumn indicate anomalous highs for total precipitable water over the Nenets Okrug on 6 November 2006 and 7 November 2013, and over YNAO on 7 November 2006 and 8 November 2013 (Fig. 13.5a–e). According to AIRS data,

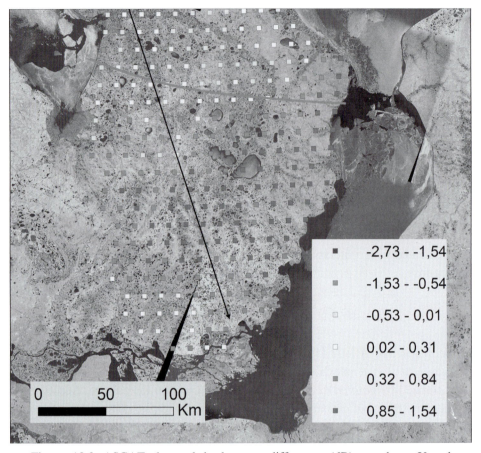

Figure 13.3 ASCAT detected backscatter difference (dB), southern Yamal Peninsula 10 November 2013. The warmer colours (yellow, orange, red) corres-pond to thicker, more impenetrable ice layers on the snow and/or ground surface. The pink line borders severely iced pasture areas; black arrow indicates reindeer herders' southward migration. [A black and white version of this figure will appear in some formats. For the colour version, please refer to the plate section.]

anomalously high air temperatures characterized both regions from 6 to 9 November 2013, followed by anomalous lows over Yamal Peninsula on 10 November 2013 (Fig. 13.5f–h). AIRS detected positive sensible heat fluxes from the surface to the atmosphere over the Barents Sea between 5–7 November 2006 and over the Kara Sea between 8–9 November 2006. Similar values were detected in 2013 over the Barents Sea between 5–8 November and over the Kara Sea on 9 November. ERA-Interim data forecasted high precipitation anomalies over NAO and YNAO from 6 to 7 November 2006 and corresponding dates in 2013 (Fig. 13.6a), coupled with wind advection from the south (Fig. 13.6b). ERA data are in close agreement with

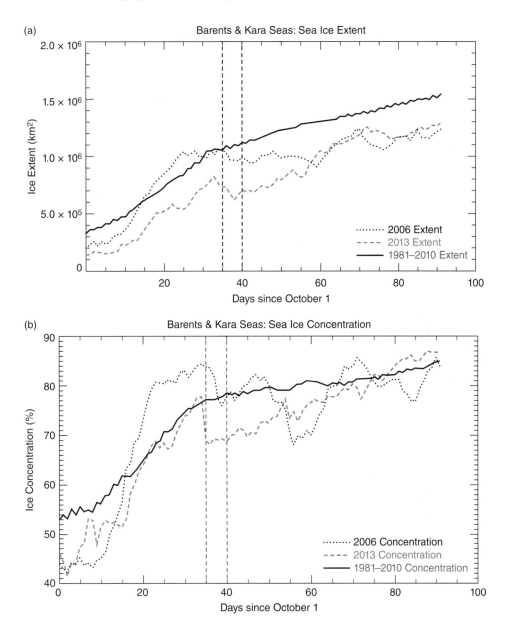

Figure 13.4 Barents and Kara Sea ice extent (top) and concentration (bottom) from SSMIS. The figures highlight between vertical dashed lines the windows of significant sea ice decrease that took place in early November 2006 (dotted line) and 2013 (dashed line). These are shown against the long-term mean of 1981–2010 (solid line).

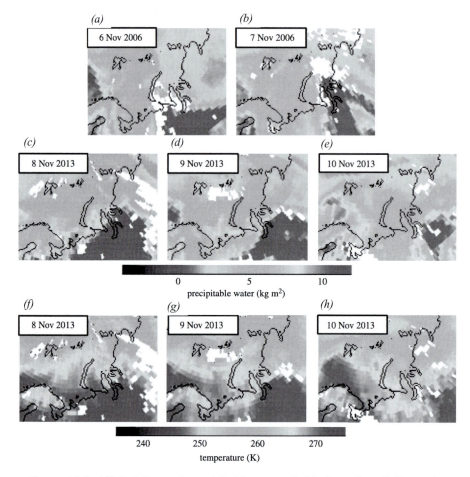

Figure 13.5 AIRS daily total precipitable water (a–b) from 6 to 7 November 2006 and (c–e) 8 to 10 November 2013 and 925 hPa temperature from (f–h) 8 to 10 November 2013 for the Barents and Kara Seas region. White indicates missing data and black outlines the coasts. [A black and white version of this figure will appear in some formats. For the colour version, please refer to the plate section.]

empirical meteorological data and incorporate sea ice data into the reanalysis modelling.

By spring–summer 2014, the private herders who had lost most or all of their animals to starvation were stranded in the tundra with no draught reindeer to move their camps. They resorted to full-time subsistence fishing and borrowed breeding stock to initiate the multi-year process of rebuilding their herds. Herders identified other historically bad icing events as occurring about once per decade since World War II, for example, in 1947, 1954, 1974 and 1996. Stammler and Ivanova (2020) describe icing events that have occurred in recent years.

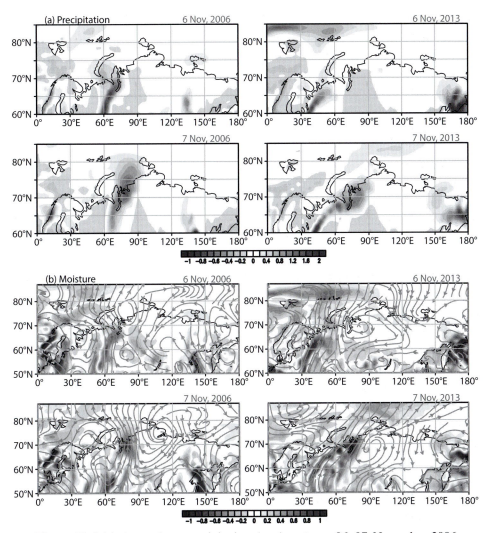

Figure 13.6 (a) Anomalous precipitation (mm) patterns 06–07 November 2006 (left column) and 06–07 November 2013 (right column). Anomalies are determined with respect to monthly averages. (b) Moisture transport (streamline) and convergence of moisture transport (shading: 10^{-7} s^{-1}) in the 1,000–850 hPa level in 2006 and 2013. Moisture convergence is overall in reasonable agreement with precipitation, suggesting that convergence of increased moisture is primarily responsible for precipitation. [A black and white version of this figure will appear in some formats. For the colour version, please refer to the plate section.]

Nenets herders sought deeper explanations for the catastrophic winter of 2013–2014 with its high reindeer mortality. Our oral history fieldwork unearthed Nenets' conceptions of such disasters as due to spiritual factors, which brought significant 'bad luck' and loss of reindeer. Two avenues of thought emerge from our preliminary analyses: (1) a breakdown of spiritual values among younger herders who tend to consider reindeer as material property for obtaining money; and (2) a breakdown of traditional rules guiding the behaviour of women in Nenets society.

With regard to point (1), one Nenets woman in her sixties criticized 'recent changes in the life of tundra people'. Specifically, she lamented the role of snowmobiles in the changing of young Nenets' attitudes about reindeer,

Before, reindeer had more spiritual value in Nenets society, but now the young generation of Nenets consider it [reindeer] as material property and for getting money.

Concerning point (2), one herder speculated that 'probably they lost this high number of reindeer because some of them [young herders] did not properly respect Nenets cosmology and traditional beliefs'. However, this opinion – while vivid – does not necessarily coincide with mainstream perspectives among actively nomadic Yamal Nenets. Most nomads seem to evaluate their present situation and difficulties with reindeer herding work more rationally, and explain openly about their real and perceived problems because of rapidly expanding industrialization of the tundra, natural gas extraction and related degradation of reindeer pastures and, finally, climate change.

Discussion

Warmer, wetter Arctic winters over the past decade have raised concerns (Ye et al., 2008; Boivert and Stroeve, 2015), particularly regarding the impacts of rain-on-snow events, on Eurasian semi-domesticated and wild reindeer populations (Bartsch et al., 2010; Hansen et al., 2011, 2014; Anonymous, 2015). Recent winter sea ice retreat is most pronounced in the Barents Sea (Årthun et al., 2012). This has been attributed to increasing 'Atlantification' of the Arctic Ocean, whereby its temperatures are becoming more like the warmer Atlantic Ocean (Smedsrud et al., 2013), and has been linked to anomalous warm advection over the Barents Sea in light ice years (Inoue et al., 2012). Another factor in Barents and Kara sea ice loss is Ural Blocking with a positive North Atlantic Oscillation, which was more persistent during the period 2000–2013 than the period 1979–1999. Ural Blocking constitutes a pattern of anti-cyclonic atmospheric circulation that affects synoptic amplification of high pressure systems between the East European north and Siberia, that is, east of the Ural Mountains. Modelling efforts, such as atmospheric re-analyses, including the ERA-Interim, have indicated anomalously high winter sensible heat flux over the Barents Sea (Lindsay et al., 2014). In a recent circumpolar comparison, the Barents Sea was the only region with significant warming in all models (Lindsay et al., 2014). Negative ice concentration anomalies were most pronounced in the Barents Sea, with significant precipitation increases over regions experiencing heightened winter sea ice loss, and atmospheric warming spreading to neighbouring landmasses around the Barents Sea (Screen et al., 2014). However, a decrease of winter sea ice cover in the Barents and Kara Seas does not always result in *a priori* expected warming over adjacent continental areas (Petoukhov and Semenov, 2010). At the same time, quantitative links between Arctic sea ice retreat and increasing precipitation remain difficult to ascertain (Wegmann et al., 2015), yet important linkages between November Barents Sea ice loss and the North Atlantic Oscillation sign for the following winter have been reported (Koenigk et al., 2016).

Meteorological data show that rain fell over the central YNAO on 7 November 2006 and on 8–9 November 2013 (Sokolov et al., 2016), but not on the forest-tundra to the south nor on the tundra zones to the north. AIRS data for winter 2013–2014 (not shown) support herders' observations that cold weather prevailed after an ice crust formed on 10 November 2013. ASCAT, despite low sensitivity to snow structural change relative to QuikSCAT (Bartsch et al., 2010), was able to capture the event owing to its intensity and severity (Fig. 13.6a). The ERA-Interim pattern of moisture convergence is similar to that of detected precipitation (Fig. 13.6b). Although wind streamline indicates that moisture is transported from the south, this does not mean that moisture originally derived from the continent but could have been introduced earlier from the sea.

Some of the individual elements in the chain of events that proceeds from warming – sea ice decline – increased precipitation and winter temperatures – increased rain-on-snow events – catastrophic reindeer mortality – social-ecological system (SES) resilience (Fig. 13.7) have been reported elsewhere (Ye et al., 2008; Bartsch et al., 2010; Bhatt et al., 2010; Hansen et al., 2011; Årthun et al., 2012; Inoue et al., 2012; Forbes, 2013; Hansen et al., 2014; Lindsay et al., 2014; Screen et al., 2014; Boisvert and Stroeve, 2015; Kim et al., 2016; Koenigk et al., 2016; Sokolov et al., 2016). However, (1) this is the first time that an entire and integrated picture is presented for a region where the coupling is so clearly manifested between sea ice-related environmental changes and critically important semi-domesticated reindeer nomad SES; and (2) this is the first study to propose a link between brief but spatially significant retreat of November sea ice and massive rain-on-snow events over the Russian mainland. We stress the idea that sea ice declines are very likely linked to increased precipitation and higher temperatures in the Barents and Kara Seas region. It is logical to infer that rain-on-snow events will be more frequent in these situations.

The pattern of more frequent and intense autumn/winter rain-on-snow events in NAO and YNAO mirrors that of summer high-pressure systems over West Siberia. Together, the regional autumn, winter and summer warming trends present major

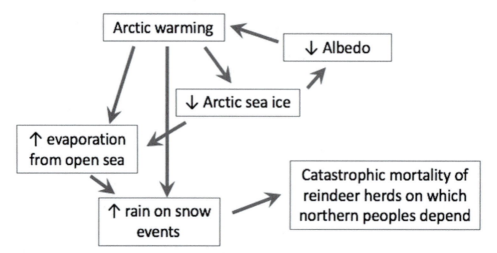

Figure 13.7 Elements in the relatively short chain of feedback processes that encompass Arctic Warming –> Sea ice decline –> Increased precipitation and winter temperatures –> rain-on-snow events –> Reindeer mortality.

challenges to maintaining tundra reindeer nomadism as a viable livelihood. Yet, Indigenous peoples have their own data and perspectives to contribute towards understanding climate change (Barnes et al., 2013; Eicken, 2013; Forbes, 2013; Savo et al., 2016) and consciously facilitate their own social-ecological resilience through collective agency (Forbes, 2013).

Nenets' oral histories reveal that smaller, more nimble, privately owned herds tend to fare better than larger collective herds in the context of long-distance, year-round migration because they can react more quickly to rapid changes in conditions. This strategy has worked well for dealing with encroaching infrastructure (Forbes, 2013). However, mortality in the context of the 2013–2014 *gololëd* event severely reduced or completely wiped out those small private herds that were virtually surrounded by areas covered with ice and unable to attain accessible pastures owing to the magnitude of the event. The location of both private and collective herds along their migration route when the *gololëd* hit affected the level of mortality. Those farthest from penetrable snow and available fodder suffered most.

Human–animal relations as well as Nenets reindeer herd sizes have always been in flux. Population fluctuation has been often explained using either natural or social drivers or a combination of both (Krupnik, 1993; Golovnev, 1995; Stammler, 2005). We are still only at the beginning of analysing social anthropological data pertaining to circumstances leading up to, during and following the *sered po* of 2013–2014 and other historically bad *gololëd* events. Local knowledge of these extreme weather events is clearly being passed from one generation to the next as elders work to ensure that traditions are adhered to. Younger herders were able to cite historic events that were experienced by their parents and grandparents.

Our preliminary findings indicate that tundra Nenets cosmology figures prominently in their interpretations of the causality behind increasingly drastic reindeer mortality episodes. These cosmological explanations may be at odds with other interpretations. For example, on the one hand, the possibility that the behaviour of women has wrought 'bad luck' and caused reindeer losses has been openly discussed among herders since 2013–2014. On the other hand, the retention of nuclear families with critical gender roles played by women is widely credited for facilitating the long-term resilience of the Yamal Nenets social-ecological system relative to, for example, the Nenets Autonomous Okrug (Golovnev and Osherenko, 1999; Stammler, 2005; Forbes et al., 2009; Forbes, 2013). Similarly, there has been internal (i.e., among Nenets) and administrative admonishment of increasingly 'greedy' private herders for keeping too many animals grazing on the tundra pastures, which subsequently contributes to high mortality rates when *gololëd* occurs. However, several of these same herders are considered to be 'successful' and actually were able to donate breeding stock to rebuild neighbouring herds after many private owners lost most or all of their animals to starvation in the *sered po* of 2013–2014. One very elderly herder commented that maybe this catastrophe 'will teach them to take their reindeer herding work more seriously'. Thus, the situation is highly nuanced. It is certainly somewhat coloured by the fact that the Yamal Nenets have been steadily modernising – with, for example, snowmobiles, mobile phones, portable generators, laptops and DVD players – at the same time as the outside world is just awakening to their role as one of the 'last' truly nomadic Arctic cultures. It is easy, but far too simplistic, for outsiders to equate the adoption of useful technological tools with 'greed' or degradation of 'traditional' cultural norms.

Finally, it must be noted that the regional warming in the tundra zone of West Siberia already exceeds the 1.5°C scenario envisioned by the COP 21 Paris agreement of 2015. Our analysis suggests that decreasing November Barents and Kara sea ice extent is linked to precipitation over coastal lands, putting Nenets herds and herders at risk. If Barents and Kara sea ice continues to decline, better forecasts of autumn ice retreat coupled with additional mobile slaughterhouses could help to buffer against losses owing to reindeer starvation following future rain-on-snow events. Even a few days of early warning could make a critical difference. This could be achieved via the combined use of empirical data on sea ice extent in real time coupled with ERA-reanalyses. Ensuring the continuity of tundra nomadism within the Arctic's largest natural gas complex under a warming climate will require meaningful consultation with herders, as well as ready access to – and careful interpretation of – real-time meteorological and sea ice data and modelling.

Acknowledgements

This research has been funded by the Academy of Finland (Decisions #256991 and #251111) and JPI Climate (#291581). Prof Forbes was generously supported by the Dickey Center for International Understanding at Dartmouth College with a visiting fellowship during 2015–2016. We are grateful to all the Yamal reindeer herders whose sincere concerns and intimate ways of engaging with the tundra not only spurred this research but were also indispensable in realizing it via their intensive participation in all seasons.

Note

1 An *okrug* is a Russian federal district.

References

Anonymous. 2015. *Report on the Socio-economic Situation of the Yamalskii District in the Year 2014 (in Russian)*. Department of Economics, Municipal Administration, Yar Sale, YNAO.

Årthun, M., Eldevik, T., Smedsrud, L. H., Skagseth, Ø. and Ingvaldensen, R. B. 2012. Quantifying the influence of Atlantic heat on Barents Sea ice variability and retreat. *Journal of Climate*, 25: 4736–4743 (https://doi.org/10.1175/JCLI-D-11-00466.1).

Barnes, J. M., Dove, M., Lahsen, M., Mathews, A., McElwee, P., McIntosh, R., Moore, F., O'Reilly, J., Orlove, B., Puri, R., Weiss, H. and Yager, K. 2013. Contribution of anthropology to the study of climate change. *Nature Climate Change*, 3: 541–544.

Bartsch, A., Kumpula, T., Forbes, B. C. and Stammler, F. 2010. Detection of snow surface thawing and refreezing using QuikSCAT: Implications for reindeer herding. *Ecological Applications*, 20: 2346–2358.

Bhatt, U. S., Walker, D. A., Raynolds, M. K., Comiso, J. C., Epstein, H. E., Jia, G., Gens, R., Pinzon, J. E., Tucker, C. J., Tweedie, C. E. and Webber, P. J. 2010. Circumpolar Arctic tundra vegetation change is linked to sea ice decline. *Earth Interactions*, 14(8): 1–20.

Boisvert, L. N. and Stroeve, J. C. 2015. The Arctic is becoming warmer and wetter as revealed by the Atmospheric Infrared Sounder. *Geophysical Research Letters*, 42: 4439–4446 (https://doi.org/10.1002/2015GL063775).

Crate, S. A., Forbes, B. C., King, L., and Kruse, J. 2010. Contact with nature. In Larsen, J. N., Schweitzer, P. and Fondahl, G. (eds.) *Arctic Social Indicators: a Follow-up to the Arctic Human Development Report*. TemaNord 2010:519. Copenhagen: Nordic Council of Ministers, pp. 109–127.

Eicken, H. 2013. Arctic sea ice needs better forecasts. *Nature*, 437: 431–433 (https://doi.org/10.1038/497431a)

Forbes, B. C. 2013. Cultural resilience of social-ecological systems in the Nenets and Yamal-Nenets Autonomous Okrugs, Russia: A focus on reindeer nomads of the tundra. *Ecology and Society*, 18(4): 36 (https://doi.org/10.5751/ES-05791-180436)

Forbes, B. C and Stammler, F. M. 2009. Arctic climate change discourse: The contrasting politics of research agendas in the West and Russia. *Polar Research*, 28: 28–42.

Forbes, B. C., Stammler, F., Kumpula, T., Meschtyb, N., Pajunen, A. and Kaarlejärvi, E. 2009. High resilience in the Yamal-Nenets social-ecological system, West Siberian Arctic, Russia. *PNAS*, 106: 22041–22048.

Golovnev, A. 1995. *Talking Cultures: Samoyed and Ugrian Traditions (in Russian)*. Ekaterinburg: Russian Academy of Sciences.

Golovnev A. V. and Osherenko, G. 1999. *Siberian Survival: The Nenets and Their Story*. Ithaca, NY: Cornell University Press.

Hansen, B. B., Aanes, R., Herfindel, I., Kohler, J. and Sæther, B.-E. 2011. Climate, icing, and wild arctic reindeer: Past relationships and future prospects. *Ecology*, 92: 1917–1923.

Hansen, B. B., Isaksen, K., Benestad, R. E., Kohler, J., Pedersen, Å. Ø., Loe, L. E., Coulson, S. J., Larsen, J. O. and Varpe, Ø. 2014. Warmer and wetter winters: Characteristics and implications of an extreme weather event in the High Arctic. *Environmental Research Letters*, 9 (https://doi.org/10.1088/1748-9326/9/11/114021).

Inoue, J., Hori, M. E. and Takaya, K. 2012. The role of Barents Sea ice in the wintertime cyclone track on emergence of a warm-Arctic cold-Siberian anomaly. *Journal of Climate*, 25: 2561–2568 (https://doi.org/10.1175/JCLI-D-11-00449.1).

Kim, K.-Y., Hamlington, B. D., Na, H. and Kim, J. 2016. Mechanism of seasonal Arctic sea ice evolution and Arctic Amplification. *The Cryosphere*, 10: 1–12 (https://doi .org/10.5194/tc-10-1-2016).

Koenigk, T. et al. 2016. Regional Arctic sea ice variations as predictor for winter climate conditions. *Climate Dynamics*, 46: 317–337 (https://doi.org/10.1007/ s00382-015-2586-19).

Krupnik, I. 1993. *Arctic Adaptations: Native Whalers and Reindeer Herders of Northern Eurasia*. Dartmouth, NH: University Press of New England.

Krupnik, I. and Jolly, D. (eds.). 2002. *The Earth Is Faster Now: Indigenous Observations of Arctic Environmental Change*. Fairbanks, AK: ARCUS.

Larsen, J. N. and Anisimov, O. 2014. Polar regions. In *Climate Change 2014: Impacts, Adaptation, and Vulnerability. Part B: Regional Aspects. Contribution of Working Group II to the Fifth Assessment Report of the Intergovernmental Panel on Climate Change*. Cambridge: Cambridge University Press, pp. 1567–1612.

Lindsay, R., Wensnahan, M., Schweiger, A. and Zhang, J. 2014. Evaluation of seven different atmospheric reanalysis products in the Arctic. *Journal of Climate*, 27: 2588–2605 (https://doi.org/10.1175/JCLI-D-13-00014.1).

Macias Fauria, M., Forbes, B. C., Zetterberg, P. and Kumpula, T. 2012. Eurasian Arctic greening reveals teleconnections and the potential for structurally novel ecosystems. *Nature Climate Change*, 2: 613–618 (https://doi.org/10.1038/NCLIMATE1558).

Moscow Times. 2014. Tens of thousands of reindeer die of extreme weather in Russia's North. *The Moscow Times*, 13 May 2014, www.themoscowtimes.com/2014/05/13/ tens-of-thousands-of-reindeer-die-of-extreme-weather-in-russias-north-a35403 (last accessed (June 26, 2020).

Petoukhov, V. and Semenov, V. A. 2010. A link between reduced Barents-Kara sea ice and cold winter extremes over northern continents. *Journal of Geophysical Research*, 115, D21111 (https://doi.org/10.1029/2009JD013568).

Post, E., Bhatt, U. S., Bitz, C. M., Brodie, J. F., Fulton, T. L., Hebblewhite, M., Kerby, J., Kutz, S. J., Stirling, I. and Walker, D. A. 2013. Ecological consequences of sea-ice decline. *Science*, 341, 519–524 (https://doi.org/10.1126/science.1235225).

Savo, V., Lepofsky, D., Benner, J. P., Kohfeld, K. E., Bailey, J., Lertzman, K. 2016. Observations of climate change among subsistence-oriented communities around the world. *Nature Climate Change*, 6 (https://doi.org/10.1038/NCLIMATE2958)

Screen, J. A., Deser, C., Simmonds, I. and Tomas, R. 2014. Atmospheric impacts of Arctic sea-ice loss, 1979–2009: Separating forced change from atmospheric internal variability. *Climate Dynamics*, 43, 333–344 (https://doi.org/10.1007/s00382-013-1830-9).

Smedsrud, L. H., Esau, I., Ingvaldsen, R. B., Eldevik, T., Haugan, P. M., Li, C., Lien, V. S., Olsen, A., Omar, A. M., Otterå, O. H., Risebrobakken, B., Sandø, A. B., Semenov, V. A. and Sorokina, S. A. 2013. The Role of the Barents Sea in the Arctic Climate System. *Reviews of Geophysics*, 51: 415–449 (https://doi.org/10.1002/rog.20017).

Sokolov, A. A., Sokolova, N. A., Ims, R. A., Ludovic Brucker, L. and Ehrich, D. 2016. Emergent rainy winter warm spells may promote boreal predator expansion into the Arctic. *Arctic*, 69: 121–129 (https://doi.org/10.14430/arctic4559).

Staalesen, A. 2016. On Arctic island, a reindeer tragedy. *The Independent Barents Observer*, 4 May 2016 (Last accessed June 26, 2020). http://thebarentsobserver .com/ecology/2016/05/arctic-island-reindeer-tragedy.

Stammler, F. 2005. *Reindeer Nomads Meet the Market: Culture, Property, and Globalization at the End of the Land*. Berlin: Lit.

Stammler, F. M. and Ivanova, A. 2020. From spirits to conspiracy? Nomadic perceptions of climate change, pandemics and disease. *Anthropology Today*, 36: 8–12.

Stroeve, J. C., Serreze, M. C., Holland, M. K., Kay, J. E., Malanik, J. and Barrett, A. P. 2012. The Arctic's rapidly shrinking sea ice cover: A research synthesis. *Climate Change*, 110, 1005–1027 (https://doi.org/10.1007/s10584-011-0101-1).

Wegmann, M., Orsolini, Y., Vasquez, M., Gimeno, L., Nieto, R., Bulygina, O., Jaiser, R., Handorf, D., Rinker, A., Dethloff, K., Sterin, A. and Brönnmann, S. 2015. Arctic moisture source for Eurasian snow cover variations in autumn. *Environmental Research Letters*, 10 (https://doi.org/10.1088/1748-9326/10/5/054015).

Ye, H., Daqing, Y. and Robinson, D. 2008. Winter rain on snow and its association with air temperature in northern Eurasia. *Hydrological Processes*, 22: 2728–2736 (https://doi.org/10.1002/hyp.7094).

14

Rising Above the Flood: Modifications in Agricultural Practices and Livelihood Systems in Central Amazonia – Perspectives from Ribeirinho and Indigenous Communities

ANGELA MAY STEWARD, RAFAEL BARBI COSTA E SANTOS, CAMILLE ROGNANT,
FERNANDA MARIA DE FREITAS VIANA, JULIA VIEIRA DA CUNHA ÁVILA,
JESSICA POLIANE GOMES DOS SANTOS, JACSON RODRIGUES AND SAMIS VIEIRA

Introduction

In the Central Amazonian white-water floodplains (*várzea*), the rhythms of life and work are defined by annual flooding patterns. During the rainy season, river waters can rise up to 12 metres annually, and residents' homes, fields and gardens come under water for several months of the year (Junk, 1997; Queiroz, 2005). During the dry season, when rivers and streams recede, they become more difficult to access. Traditional farmers and Indigenous peoples who live along the banks of streams and rivers have long adapted their daily lives, agricultural practices and natural resource management systems in concert with annual flood patterns. Environmental dynamism coupled with ever-changing social and economic dynamics have stimulated traits of adaptability and resilience which have been central to the survival and social reproduction of the residents over time (Harris, 1998; Adams et al., 2009).

While stark seasonality is characteristic of the region, and in this sense floods are nothing new, changes in flooding patterns in recent years have been observed and described by floodplain farmers of the Boca de Mamirauá community, Amazonas state, Brazil. As one female resident from this community stated during her interview, 'Today large floods are more frequent; they always come larger. There is no time to harvest our crops. The flood comes and kills [everything] ... '

There is a general consensus among rural residents that severe floods, previously described as anomalies, are more frequent and last longer, as expressed by a male farmer from the same community, 'The 2009 flood almost flooded the entire house. Another large one came in 2012. This flood [of 2015] seemed to last longer, well, it was very large and took a long time to recede ... ' These observations are corroborated by research on regional climatic patterns which describe an increase in the frequency of extreme hydrological events, floods and droughts, many of which are classified as 'once in a century' events in the Amazon

Basin, but which have occurred frequently during the last two decades (Marengo and Espinoza, 2015). These events are, in part, attributed to El Niño and La Niña events but also to irregularities in moisture transport; in the case of severe floods, a recent study links the strengthening of the Walker circulation (i.e., east–west atmospheric circulation cells along the Equator; Lau and Yang, 2015) to the observed increase in frequency and intensity of severe floods in the Amazon. Circulation change is considered to be part of 'a tropical-wide climate reorganization triggered most likely by rapid tropical Atlantic warming during recent decades' (Barichivich et al., 2018, p. 4). Scientists predict that climate and land use changes in the region will increasingly make hydrological conditions more unpredictable and likely increase severe floods and droughts in future years. Faced with this scenario, they emphasize the importance of studies that better our understanding of how natural and human systems cope with extreme environmental changes. Community-level research may help mitigate the impacts of such events, especially on vulnerable peoples who depend on agricultural and fishing activities for their survival (Marengo et al., 2012; Marengo and Espinoza, 2015).

In line with this reasoning, the current chapter discusses the challenges faced by traditional and Indigenous farmers in Central Amazonia owing to the increasing frequency of extreme hydrological events. Here, we focus on the impacts of large floods while acknowledging that extreme droughts are simultaneously occurring in the region. We specifically document the immediate responses of farmers to extreme floods in 2009 and 2015, and how residents are modifying their agricultural systems and natural resource management practices. Farmers' testimonies, recorded in the field by the authors of this chapter, guide this text and join the voices of other Indigenous and traditional peoples in this collection whose lives and livelihoods are changing as a result of the climate crisis.

Background

Place and People

Research was conducted in the middle Solimões region, Amazonas state, Brazil and involved ten communities located within the Mamirauá and Amanã Sustainable Development Reserves (Fig. 14.1). The Mamirauá reserve is located at the confluence of the Solimões and Japurá rivers, with its easternmost region located close to the city of Tefé. This reserve is unique in strictly protecting floodplain forests (Queiroz and Peralta, 2006). The Amanã reserve encompasses forested areas in upland, floodplain and *paleovárzea* environments; areas of flooded black water forests (*igápo* in Portuguese) are also found, contributing to a local pattern of biodiversity (Queiroz, 2005; Irion et al., 2011). *Várzea* areas, as

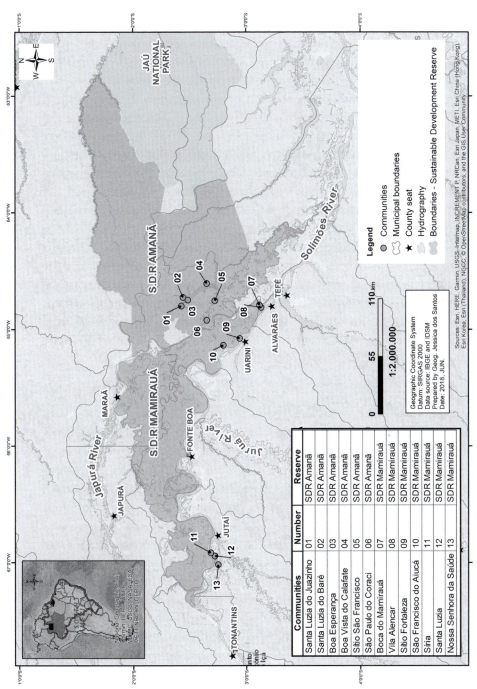

Figure 14.1 Map showing the location of study communities in Mamirauá and Amanã Sustainable Development Reserves, Amazonas, Brazil. [A black and white version of this figure will appear in some formats. For the colour version, please refer to the plate section.]

Map credit: Jessica Poliane Gomes dos Santos

235

they are known in Portuguese, are subject to annual flooding, while upland areas escape flooding, being slightly higher in elevation. *Paleovárzea* areas are intermediate zones that pertain to older geological formations consisting of old alluvial deposits. Parts of these areas are inundated during high flood years (Irion et al., 2011). Locally, residents refer to areas of *paleovárzea* as *terra firme* or uplands; we maintain this denomination for the remainder of this text.

The Mamirauá reserve hosts a population of approximately 11,532 residents distributed over 207 localities; while 3860 people, distributed among 86 communities, reside in the Amanã reserve (Steward, 2018). Reserve residents largely belong to the rural Amazonian peasantry and are of a mixed cultural background of Indigenous, European and African descent. These groups are referred to as *ribeirinhos* (literally 'riverbank dwellers' in Portuguese), or sometimes as *caboclos*, and emerged as a social group following Portuguese colonization (Adams et al., 2009).

Mamirauá is also home to the Cocama, Ticuna, Miranha and Omágua Indigenous groups who also sometimes identify themselves as *ribeirinhos*, since this term refers to a social category as well as to a dwelling place, that is, the floodplain. Amanã is also home to the Miranha and Mura Indigenous groups (Lima et al., 2010). This chapter includes observations and insights from *ribeirinho* groups in both reserves and from Cocama groups of the Upper Solimões region of the Mamirauá reserve (Fig. 14.1).

Livelihood Systems and Agricultural Production

Mamirauá and Amanã residents practice swidden fallow agriculture and, as is typical in the region, manioc or cassava (*Maniot esculenta* L.) is the primary crop. The importance of manioc is derived from its key place in the regional diet, which is based in manioc flour and fish (Adams et al., 2009; Lima et al., 2012). Swidden farming is combined with fishing, hunting, and timber and non-timber forest extraction. More recently, scholars have discussed the role of non-rural incomes and wages in livelihood systems in the middle Solimões region (Peralta and Lima, 2014).

Agricultural practices in Mamirauá and Amanã reserves include the general steps of clearing a forested area, burning felled vegetation and further clearing, followed by planting, weeding and harvesting, with variations occurring between main environmental areas: floodplain/*várzea* and upland/*terra firme* areas. Floodplain farming activities are carried out in line with seasonal flood patterns, being confined to the periods of *vazante* (draining of the basin) and *seca* (drought or dry period). Here, clearing and burning are carried out from July to September and planting is done from September to October. Harvesting takes place from

April to May when flood waters begin to rise. As floodplain fields are harvested, manioc tubers are transported to processing areas where families grate the tubers, sieve out *tucupi* (bitter juice) and toast the mass to produce manioc flour (Steward, 2018). Since manioc is planted with stems – referred to as *manivas*, this planting material is either transferred to higher ground, where it is maintained in the soil as a living seed bank, or cut and stored on floating platforms or other structures to withstand the flood.

When flood waters recede in June, temporary landforms such as mud flats and beaches appear. Farmers use these areas to plant short-cycle crops such as corn (*Zea mays*), beans (*Phaseolus vulgaris*), squash (*Cucurbita* sp.), melon (*Cucumis* sp.) and watermelon (*Citrullus lanatus*). Forested and fallow areas located on slightly higher ground as well as areas surrounding residences are planted with more permanent crops such as fruits, which are more resistant to constant flooding; taken together, these areas provide families with a great diversity of crops.

Floodplain production systems are intricately timed with the annual rise and fall of river waters. Thus, changes to the flooding regime greatly impact agricultural practices and production yields. Upland areas which escape floods, on the other hand, present fewer production challenges. Tubers can remain in the soil for up to two years and are harvested intermittently according to household needs throughout the year. In and around upland communities in the Amanã reserve, farmers maintain diverse home gardens and other traditional agroforest systems (*sítios*), earning income from the sale of various fruits.

Methods

Data collection was carried out by researchers and technicians from the Research Group for Amazonian Agriculture at the Mamirauá Sustainable Development Institute (IDSM) as well as members of the Social Organization research group who specifically studied the response of Cocama groups to extreme floods. In the latter case, data were collected as part of ongoing ethnographic research on the Cocama, their social patterns and historical occupation in the region (see Lima et al., 2010 and Santos, 2014).

The Research Group for Amazonian Agriculture works collaboratively with farmers in the reserves to develop projects in sustainable agriculture. The team observed farmers' adaptive responses to extreme floods during regular monthly visits to support active projects and recorded them in monthly reports. When the 2015 flood waters receded, team members conducted seventeen semi-structured interviews following Bernard (2011): ten in floodplain communities and seven in upland communities. Farmers were specifically asked to describe the nature of the flood and how it compared to previous events; impacts on agricultural areas, including seed and germplasm losses; other investment losses such as houses and livestock; and social and economic impacts. We also asked residents to describe their immediate reactions and responses to the event and long-term planning to prepare for future events.

Results

Impacts of Extreme Floods and Residents' Responses: Várzea Communities

Cocama of the Upper Solimões region

During ethnographic research in the communities of Síria, Santa Luzia and Nossa Senhora da Saúde in March of 2010, families described themselves as being in the process of recovering from the 2009 flood. Having lost most of their manioc fields and seed stocks, they were purchasing manioc flour from local urban markets at an inflated cost. Families reportedly lost at least seven manioc varieties, both sweet and bitter; and as one resident from the community of Nossa Senhora da Saúde stated, 'These varieties are truly lost, you can't find them anymore anywhere in the *várzea*.'

This statement is all the more significant since floodplain manioc varieties are adapted to local conditions and selected because their tuberous roots develop quickly, allowing farmers to harvest fields before they are inundated by annual floods. When residents acquire *manivas* from upland fields after heavy floods, plants go through a process of adaptation to the floodplain. Farmers report that first-year yields from such plants are quite low, leading to shortages in food supplies. This was evident in 2010, when Cocama residents in the process of harvesting manioc reported that yields would not be sufficient to produce enough manioc flour to sustain their families until the next growing season.

In addition to losing *maniva* varieties, Cocama farmers also reported losing six different fruit species from their orchards (Table 14.1). Even some plants considered resistant to floods, such as *ingá* (*Inga* sp.), did not survive. Following such losses and owing to the importance of cultivation to Cocama society who equate having diverse cultivated spaces with abundance and food security, some residents have recently begun to plant in upland villages, especially in the Indigenous land, Estrela da Paz (village of Issapó). Others have reportedly acquired small lots in upland areas of the city of Fonte Boa to continue planting manioc and guarantee their supply of seed material.

Ribeirinho *Farmers, Mamirauá Reserve*

The ten *ribeirinho* farmers interviewed in the Mamirauá reserve agreed that the 2015 flood was the largest (*maior* in Portuguese) they had seen in their lifetimes; its 'greatness' was attributed to the enormous volume of water, the rapidity with

Table 14.1 *Summary of reported impacts of extreme flood events (years 2009 and 2015), Mamirauá and Amanã Sustainable Development Reserves, Amazonas State, Brazil*

Study area/ social group/ Flood year	Impacts on agricultural production	Species and varieties lost in the floods	Other impacts
Mamirauá reserve/ Cocama communities/ 2009 flood	Loss of manioc fields and agroforests; partial relocation of manioc fields to upland areas.	Lemon, lime and oranges, *araça* (*Eugenia stipitata*) *cupuaçu* (*Theobroma grandiflorum*) and *abiu* (*Pouteria caimito*).	Damage to houses; food insecurity; increase in manioc flour prices; temporary relocation of families to urban areas
Mamirauá reserve/ *Ribeirinho* communities/ 2015 flood	Loss of manioc fields and agroforests; partial relocation of manioc fields to upland areas.	Oranges, lime, lemon, and *açaí* (*Euterpe oleracea*), bananas, squash, miscellaneous herbs.	Damage to houses; closing of churches and schools; temporary relocation of families to upland areas; decrease in quality of life during flood months; food insecurity.
Amanã reserve/ *Ribeirinho* communities/ 2015 flood	Partial losses of manioc fields and agroforests.	Fruit species: pineapple (*Ananas comosus*), avocado (*Persea americana*), oranges, *cupuaçú, abiu, mari* (*Cassia leiandra*), *ingá, açaí.* *Maniva* varieties: *Tapaiona, sacai, catombo, sete ano, baixotinha, macaxeira*	Loss of livestock; diminished quality of life; transportation difficulties during flood months.

which the river rose and how slowly it receded. A female floodplain farmer from Vila Alencar, Mamirauá Reserve recounted,

Our entire house was covered in the 2015 flood. One night during very heavy rains, we had to move to my father's house after I realized that the house was going under. The flood came quicker this year…it came all at once and stopped, such that the time for agricultural production was cut short. Everybody just kept looking at the [rising] waters. The flood was longer this year; the ground finally appeared in September, when it should have appeared during the month of July.

In the communities of the Boca de Mamirauá and Vila Alencar, all manioc fields were entirely inundated. Citrus fruit trees such as oranges, lime, lemon and *açaí* (*Euterpe olearacea*) in agroforests, planted on higher levees in the floodplain and around residences, were also destroyed. Farmers reported losing the following seed material: bitter and sweet manioc varieties, banana (*Musa* sp.) stems, watermelon, squash and diverse herb seeds (Table 14.1). These losses were attributed to the fact that not only were flood waters higher, but they rose faster and earlier than in normal years. This did not give residents sufficient time to harvest their manioc fields and save seed materials. It is important to note that some losses are temporary, spanning the course of one growing season, until farmers can recover seed stocks. Others may be permanent, when farmers are not able to recover seed stocks for specific species or varieties owing to general regional losses.

Moreover, when the waters were rising, residents tried to save various animals by placing them on floating rafts or penning them inside their homes. Despite these efforts, many reported animal losses including cattle, chickens and pigs, as well as dogs, which are not only pets but also hunting companions. By November, eight of the ten families had run out of their own manioc flour and were purchasing it from regional cities.

When flood waters rise, floodplain families either remain in their communities or leave to temporarily live in nearby cities located in upland areas. Families who remain do so because they do not have a place to stay in the city or do not want to abandon their property in the floodplain, fearing theft or further destruction. Families who remain in their homes during floods slowly raise the wooden floorboards with rising waters. Thus, during the flood months, families live in cramped conditions between the floor and ceiling. Owing to the severity of the 2015 flood, families reported feeling 'squashed' since little space remained between the floor and ceiling at the peak of the flood.

Other reported challenges included difficulties in accessing fresh drinking water, and health conditions related to poor water quality. During the flood, fishing was also difficult, as fish dispersed throughout the basin with the high waters. They further discussed the dangers of living in close proximity to *jacarés* (alligators) and poisonous snakes that came with the flood waters, which were particular concerns for households with small children. Three floodplain households reported losing livestock to alligators.

On the other hand, families who left for the city described situations of overcrowding at relatives' homes. As time wore on, many felt that they were becoming a burden to their families since all food in the city needs to be purchased. Regardless of whether or not families temporarily relocated out of the floodplain, all households lost goods such as appliances – stoves and freezers – investment items for rural Amazonians. All families also had to re-build their homes partially or fully as a result of flood damage.

Social consequences were reported as well. Schools were closed in the region for up to three months and there was difficulty in maintaining religious meetings and parties in these times of despair. Directly after the floods, a general anxiety regarding the unpredictability of climate events remained.

Despite the gravity of the situation, positive responses also took place. During the flood itself, some farmers were able to save *manivas* by placing them on rafts (*flutuantes*) to be planted during the next growing season. The same was done for some small herbs and banana seedlings, where farmers saved selected individuals by planting them on floating structures to replant after flood waters had receded.

Farmers in Vila Alencar also stated that since 2012, they have been building on-site platforms in their manioc fields to store *manivas* above floodwaters. For short-cycle crops, farmers now store seeds in closed soda bottles, as opposed to open tins, for greater protection from humidity. There has also been a move towards different types of production, including a focus on short-cycle crops such as corn, squash and watermelon, which can be harvested quickly and immediately and contribute to household food security. Farmers also expressed an interest in investing in raised platforms and/or using old canoes to intensify the production of *hortaliças* (greens and herbs).

Like the Cocama, *ribeirinho* farmers from the Boca de Mamirauá and Vila Alencar communities discussed the need to obtain a small plot of land in an upland area. Some mentioned this as a way to maintain manioc seed stocks, while others plan to entirely move manioc production to upland areas and abandon floodplain farming. For residents of Vila Alencar and Boca de Mamirauá, the outskirts of the city of Alvarães is an important area both for gathering *manivas* after flood events and for establishing new manioc fields.

Residents of São Francisco de Aiucá also discussed moving manioc production to upland areas. Historically, manioc fields in this community were placed along the banks of the Solimões river; however, in recent years, the banks have receded or collapsed entirely as a result of erosion caused by the flooding. In general, farmers here have downsized production along riverbanks, if not abandoned cultivation altogether, moving production sites to peri-urban areas around the city of Uarani. Residents here have strong ties to the city through kinship and affinity networks and acquire parcels of land and planting materials through these networks. This was also observed in the community of Sítio Fortaleza, where families use both the upland areas around this city and low-lying areas in their communities.

Upland Communities

Ribeirinho *Farmers, Amanã Reserve*

Just as in the floodplain areas, farmers in the *paleovárzea* environments of Amanã were surprised by the nature of the 2015 flood and agree that floods are becoming more extreme over time. This sentiment was perhaps more pronounced in Amanã since, in normal years, agricultural areas do not come under water.

Of the eleven interviewed farmers, all reported that they had lost a portion of their production areas, including both manioc fields and agroforests. Losses included eight fruit species and seven manioc varieties located in the lowest lying areas of their lands (Table 14.1). However, unlike in floodplain areas, complete losses were not reported. Most farmers were able to save enough *manivas* to re-plant their fields at the end of 2015. They were also able to recover specific manioc varieties by obtaining cuttings from neighbours and relatives. All families still had sufficient manioc flour for their own consumption and all but one had enough to continue to sell manioc flour even after the flood.

In and around Amanã lake, farmers cited losing cacao (*Theobroma cacao*) plantings, which had been established when the communities were founded by their current inhabitants in the 1960s and 1970s. One male farmer from the Boa Esperança community, Amanã Reserve, noted that ' ... even *várzea* plants that are used to water died [in this year's flood]'. Another farmer reported that, in the forest, 'all wild plants also died, the *cupuí* [*Theobroma* sp.], these types of things ... '

Farmers had a good understanding as to why plants were dying and attributed it to the changing nature of floods in the region. A female farmer from the same community stated, 'I never saw cacao die in water! I think this happened here because in the floodplain water flows, correct? Here it stagnated and heated up, this is what killed the plants ... ' It is interesting to note that floodplain residents reported losses of similar fruit species in forested areas and also attributed these losses to warm stagnant waters.

Just as in the floodplain, residents lost various animals such as goats, chickens, cows and dogs. They also faced difficulties with transportation in the months that the community was largely under water. Furthermore, when floodwaters receded, they left behind piles of detritus scattered throughout residential and production areas. During the pronounced dry season that followed the 2015 flooding events, large tracts of agroforests surrounding the community of Boa Esperança, the largest community of the Lake Amanã region, caught fire and were devastated. Fires, intentionally set as part of the process of establishing new manioc fields, escaped out of control and spread quickly owing to the fodder left by the flood. The loss of agroforests was significant for the community which specializes in the sale of fruits; it also had consequences for other farmers in the region who seek out Boa Esperança farmers for seed materials.

Discussion

Research results demonstrate the impacts of the extreme floods of 2009 and 2015 and farmers' responses to these events. Interviews with farmers allowed us to understand their experiences during the flood events, their medium-term impacts and how farmers are adapting their longer-term production strategies as extreme floods occur more frequently in the region. For families who remain in rural areas, the flood months bring about conditions of scarcity, such as difficulties accessing water and fish, along with the absence of a normal social life. For those who leave for the city, they face the challenges of securing food in a place where everything needs to be purchased. Material losses are also significant for rural peoples, where household goods, such as appliances and livestock, are investment items acquired slowly over time (Lima, 2009). As expected, lower-lying floodplain communities suffered greater production and general household losses.

Farmers in all communities specified plant losses. In the case of the Cocama, specific manioc varieties were completely lost. Residents also described their immediate efforts to secure germplasm by planting herbs, manioc stems and bananas on rafts and other floating platforms during the flood. Saving these varieties allowed them to sow new areas once river waters receded. Access to *manivas* found in upland manioc fields and acquired through kinship and affinity networks was also essential to re-establishing new fields following the 2009 and 2015 floods. Regionally, *ribeirinho* farmers customarily obtain *manivas* from upland areas following large floods: a practice that occurs to a lesser degree during normal flood years (Lima et al., 2012). As is also common in the region, floodplain residents acquire seeds by means of the gift economy or reciprocal relationships. In *ribeirinho* and Indigenous societies in the middle Solimões region, certain goods such as *manivas* and other planting materials as well as fish and fruits for household consumption are exchanged on the basis of reciprocity (Harris, 2000; Lima, 2009). While Lima et al. (2012) note sporadic incidents when *manivas* were sold and purchased, this is an exception, and in our work regarding large flooding events, no farmers reported paying for seed material to re-establish their fields.

In the Amanã reserve, documented losses to agroforests and their recovery will be slow since this involves the establishment of slower-growing perennials; however, the willingness of family and neighbours to share and exchange planting material facilitates this process.

Our data point to adaptive changes provoked by the floods, including new seed storage mechanisms both for *manivas* and for annual, short-cycle crops. Perhaps more significantly, research results indicate two changes to the overall organization and structure of production systems. The first is the tendency to focus on short-cycle crops directly following a large flooding event: crops such as corn, beans, squash, melon and watermelon develop in about three months, as opposed to manioc that requires six to twelve months depending on the variety. This strategy thus helps by securing household food and income more quickly. Second, we observed farmers' efforts to acquire land in the upland areas, establishing manioc fields as seed banks or entirely moving their fields to these regions. Again, these areas are generally obtained through kin and affinity networks, which provide a basis for resilience of communities in this time of change. However, this scenario also brings to light vulnerabilities since not all families have access to upland areas for planting. In fact, some families reported that they do not have friends or kin with whom they can take refuge during the flood. During an interview in the Mamirauá reserve, one farmer told us that he was lucky to have family in an upland community with whom he could live but expressed concern that others were not so fortunate. He also discussed the added difficulties of those who are left behind; because they are few in number, the families that remain behind do not have the support networks that normally operate when the rest of the community is present. Ongoing research is needed to determine if there will, in fact, be a gradual movement away from farming in the floodplain. If this does occur as a result of increasing climate variability, it would represent a shift in resource management and in the type of relationship that people have with their environment. This is significant given that peoples' identity (*ribeirinhos* = riverbank dwellers) has long been associated with the floodplain.

Families in the study area also display a great capacity for and willingness to modify their production systems in light of new changes. While the development of new seed storage and production strategies described in this chapter displays these attributes on an individual level, information on regional manioc prices reveals the adaptive capacity of these systems at a broader level. During the 2015 flood, regional manioc flour prices remained more stable when compared to the effects of the 2009 and 2011 floods. One farmer explained that owing to manioc flour shortages in 2010 and 2012, upland farmers in the region expanded their fields and increased production, such that after the 2015 flood, prices remained stable in the cities. As such, supplies were sufficient to sustain floodplain families who were without their own manioc flour. Although these families still had to purchase manioc flour, this foresight was nonetheless beneficial to all.

Conclusion

The intent of this short discussion was to provide insights into the challenges faced by traditional and Indigenous communities in Central Amazonia as they contend with extreme flood events. As residents in Central Amazonia have always contended with regional flood patterns resulting from the seasonal rise and fall of river waters, changes in these patterns are often overlooked by policy makers who assume that communities are equipped to handle them on their own. However, the increasing frequency of extreme weather events has meant that residents need to respond to new events while they are still recovering from previous disasters. In the case of the Cocama, ongoing fieldwork has revealed that many families are becoming disheartened because their ability to enrich their dwelling places with cultivated plants, as they traditionally prefer to do, has been greatly undermined in recent years. Thus, despite evidence of the adaptive capacity of the farmers interviewed in this study, if extreme events continue to occur, it may undermine their resilience and provoke further migration of both farming areas and residences to surrounding uplands. In the context of the increasing frequency of extreme weather events, it is imperative that we understand people's experiences to better provide support and to plan for future events and the difficulties that they may cause.

As we move forward, key questions regarding the white-water floodplains remain: What is the fate of floodplain production systems in Central Amazonia? While families will likely continue to depend on floodplain areas for natural resource management, in particular for fishing, will these areas still be home to communities and their agricultural areas? This question is critical considering the importance of floodplains to *ribeirinho* identity and to the history of the human occupation of Amazonia, where the floodplain has been the site of continual occupation since Pre-Colombian times.

References

Adams, C., Harris, M. and Murrieta, R. 2009. *Amazon Peasant Societies in a Changing Environment*. Netherlands: Springer. https://doi.org/10.1007/978-1-4020-9283-1

Barichivich, J., Gloor, E., Peylin, P., Brienen, R. J., Schöngart, J., Espinoza, J. C. and Pattnayak, K. C. 2018. Recent intensification of Amazon flooding extremes driven by strengthened Walker circulation. *Science Advances*, 4(9): eaat8785. https://doi.org/10.1126/sciadv.aat8785

Bernard, H. R. 2011. *Research Methods in Anthropology: Qualitative and Quantitative Approaches*. Lanham, MD: Rowman Altamira.

Harris, M. 1998. The rhythm of life on the Amazon floodplain: Seasonality and sociality in a riverine village. *Journal of the Royal Anthropological Institute*, 4(1): 65–82. https://doi.org/10.2307/3034428

Harris, M. 2000. *Life on the Amazon: The Anthropology of a Brazilian Peasant Village*. London: British Academy Publishing.

Irion, G., Mello, J. A. S. N., Morais, J., Piedade, M. T. F., Junk, W. and Garming, L. 2011. Development of the Amazon valley during the middle to late quaternary: Sedimentological and climatological observations. In Junk, W., Piedade, M. T. F., Wittman, F., Schöngart, J. and Parolin, P. (eds.) *Amazonian Floodplain Forests*. Springer, Netherlands, pp. 27–42.

Junk, W. 1997. *The Central Amazon Floodplain: Ecology of a Pulsing System*. Springer, Netherlands. https://doi.org/10.1007/978-3-662-03416-3_1

Lau, K. M. and Yang, S. 2015. Walker Circulation. In North, G. R., Pyle, J. and Zhang, F. (eds.) *Encyclopedia of Atmospheric Sciences*, Second Edition. London: Elsevier, pp. 177–181.

Lima, D. 2009. The domestic economy in Mamirauá, Tefé, Amazonas State. In Adams, C., Murrieta, R., Neves, W. and Harris, M. (eds.) *Amazon Peasant Societies in a*

Changing Environment. Netherlands: Springer, pp. 131–156. https://doi.org/10.1007/978-1-4020-9283-1_7

Lima, D., Souza, M. O. and Santos, R. B. C. 2010. Organizações Indígenas e as Políticas de Reconhecimento no Médio Solimões. In Ricardo, C. A. and Ricardo, F. (eds.) *Povos Indígenas no Brasil 2006/2010*. São Paulo: Instituto Socioambiental, pp. 349–352.

Lima, D., Steward, A. M. and Richers, B. T. 2012. Trocas, Experimentações e Preferências: Um Estudo Sobre a Dinâmica da Diversidade da Mandioca no Médio Solimões, Amazonas. *Boletim do Museu Paraense Emílio Goeldi: Ciências Humanas*, 7(2): 371–396.

Marengo, J. A., Tomasella, J., Soares, W. R., Alves, L. M. and Nobre, C. A. 2012. Extreme climatic events in the Amazon Basin. *Theoretical and Applied Climatology*, 107(1–2): 73–85. https://doi.org/10.1007/s00704–011-0465-1

Marengo, J. A. and Espinoza, J. C. 2015. Extreme seasonal droughts and floods in Amazonia: Causes, trends and impacts. *International Journal of Climatology*, 36(3): 1033–1050. https://doi.org/10.1002/joc.4420

Peralta, N. and Lima, D. 2014. A comprehensive overview of the domestic economy in Mamirauá and Amanã in 2010. *Uakari*, 9(2): 33–62. http://dx.doi.org/10.31420/uakari.v9i2.155

Queiroz, H. L. 2005. A Reserva de Desenvolvimento Sustentável Mamirauá. *Estudos Avançados,* 19(54): 183–203.

Queiroz, H. L. and Peralta, N. 2006. Reserva de desenvolvimento sustentável: Manejo integrado dos recursos naturais e gestão participativa. In Garay, I. and Becker, B. K. (eds.) *Dimensões Humanas da Biodiversidade*. Petrópolis: Editora Vozes, pp. 447–476.

Santos, R. B. C. 2014. Direito: Posse e Maestria entre os Cocama da Foz do Jutaí/AM. Annals of the 29th Meeting of Brazilian Anthropology.

Steward, A. M. 2018. Fire use among swidden farmers in central Amazonia: Reflections on practice and conservation policies. In Fowler, C. and Welch, J. R. (eds.) *Fire Otherwise: Ethnobiology of Burning for a Changing World*. Salt Lake City: University of Utah Press, pp. 104–127.

15

Indigenous Storytelling and Climate Change Adaptation

ÁLVARO FERNÁNDEZ-LLAMAZARES AND MAR CABEZA

Introduction

The intricate relationship between Indigenous peoples and their surrounding environments has resulted in first-hand, complex and detailed bodies of Indigenous and local knowledge (hereinafter referred to as 'ILK') that have been pivotal in allowing societies to subsist in a wide range of environments (Nakashima et al., 2012). Such knowledge, which is the result and continues to be the source of adaptive strategies to changing social-ecological conditions, is key to strengthening community resilience to respond to the multiple stressors of global climate change and to deal with disturbances under conditions of high uncertainty (Fernández-Llamazares et al., 2015a, 2015b).

However, while Indigenous practices have been shown to promote social-ecological resilience over time, some of these are being compromised (Kesavan and Swaminathan, 2006; Mercer et al., 2007). For instance, in Arctic communities, cultural change is eroding the cycle of ILK transmission upon which adaptive capacity is built (e.g., Ford et al., 2010). Some studies have shown that there is a global disappearance of the traditional institutions for ILK intergenerational transmission (Iseke and Moore, 2011; Fernández-Llamazares et al., 2015a), often owing to the decreasing interaction between elders and younger generations (e.g., Herman-Mercer et al., 2016). Despite the fact that resourcefulness and adaptability are essential hallmarks of any Indigenous culture, owing to this disruption, Indigenous peoples are facing increasing challenges in adapting to rapid climatic changes.

At the same time, research has empirically shown that top-down climate change adaptation strategies targeting Indigenous peoples have often failed to present climate change information in a locally salient and credible way, tailored to the cultural values, local knowledge and/or belief systems of Indigenous peoples (Burman, Chapter 16). Climate change adaptation plans launched by government agencies and non-governmental organizations among Indigenous peoples often involve single one-off workshops that present scientific knowledge, rarely take into account local understandings of climate change and often overlook traditional mechanisms for knowledge transmission. Communication about climate change is usually too general or decontextualized from local realities and Indigenous ways of knowing, thus failing to bridge the epistemological gap between different knowledge systems stemming from different worldviews, ontologies, logics and/ or mental models (Fernández-Llamazares et al., 2015b). As a result, pathways to community adaptation involving the imposition of Western scientific concepts have generally received low local support and reduced long-term commitment (Janif et al., 2016).

In view of this, several policy instruments are calling for greater engagement of Indigenous peoples in the development of adaptation plans, directly incorporating considerations of ILK systems into these strategies (Nakashima et al., 2012). Such approaches emphasize the importance of maintaining an intergenerational collective memory to facilitate the continuous adaptation of Indigenous peoples to new social-ecological contexts, particularly in view of rapid climatic changes (Davidson-Hunt and Berkes, 2003; Iseke and Moore, 2011). Yet, as the functioning of collective longitudinal information transfer of ILK systems is disrupted, Indigenous peoples become increasingly vulnerable to the effects of climate change (Ford et al., 2010; Nakashima et al., 2012). The failure to detect, understand and interpret environmental changes has been implicated among the main barriers to successful climate change adaptation (Fernández-Llamazares et al., 2015a, 2015b). Hence, it becomes crucial to develop climate change communication plans that are tailored to the epistemic contexts of Indigenous peoples in order to ensure that they receive pertinent messages linked to their cultural values and ILK systems.

Indigenous Storytelling

Indigenous stories are preserved as a collective treasured experience and have been handed down over many generations, thus playing an important role in revitalizing Indigenous cultures in the face of ever-encroaching environmental changes (Sakakibara, 2008; Ryan, 2015). Storytelling among Indigenous peoples serves a number of goals, including entertainment, the transmission of ILK, and the maintenance of cultural identity and sense of community, to cite just a few (Lawrence and Paige, 2016). A characteristic feature of Indigenous stories is that they enable intergenerational communication (Fig. 15.1), allowing for adaptation to many different environments (Iseke and Moore, 2011). Indigenous storytelling does not just seek to represent things that have happened, but as Lawrence and

Figure 15.1 The Maasai Indigenous peoples of the Kenyan Rift Valley have a rich tradition of oral storytelling as a way of transmitting ILK across generations. Maasai oral histories are vital to the transmission of certain ecological management practices that can be crucial for climate change adaptation and conservation. [A black and white version of this figure will appear in some formats. For the colour version, please refer to the plate section.]
Photograph by Joan De La Malla

Paige (2016, p. 65) perceptively state, to link 'the past and present to a wiser future'.

It is important to note that Indigenous stories disclose conceptualizations of human-nature relations that often differ substantially from Western worldviews. In general, Indigenous stories acknowledge the intrinsic values of nature, in contrast to the instrumental values prevalent in the Western view, and transmit notions of spirituality (Lawrence and Paige, 2016). Nanson (2011, p. 140) suggests that Indigenous homelands should be conceived as 'mythscapes of stories and spirits' and that the present-day disconnection of stories and landscapes undermines our ability to mediate health relations with the natural world. For example, Yupik and Cupik storytelling in subarctic Alaska features tales of famine and harvest disruption as a direct result of anthropogenic action (Herman-Mercer et al., 2016). The customary Yupik and Cupik worldview, as transmitted by oral histories, reveals a reciprocal relationship between humans and the environment. As such, harvest disruptions or natural disasters are not perceived as externally imposed events but rather as a consequence of a transgression of the culturally established code of conduct with regard to the environment (Fienup-Riordan, 1986).

Although representations of nature vary substantially between and even within Indigenous groups, a common characteristic of many Indigenous storytelling practices is that nature generally has a voice and communication with non-human beings is part of everyday life. In most Indigenous stories, the health of society and that of the environment are intertwined and essentially inseparable. For example, the Nahua Indigenous peoples of Mexico and El Salvador share stories about hostile water-dwelling spirits watching over bodies of water whom people need to respect in order to secure water availability for the future.

While the potential of Indigenous storytelling as a cultural tool towards healthier behaviour (Hodge et al., 2002) or improved schooling (Eder, 2007) has largely been examined, there is limited research on how this time-honoured practice could make interventions to support climate change adaptation more culturally appropriate and locally meaningful. In the following sections, we present a number of case studies from all over the world showing potential applications of Indigenous storytelling for, (1) revitalising ILK systems to cope with climate change; (2) fostering a sense of place in a rapidly changing environment; and (3) making climate change communication programmes more culturally sensitive (Fig. 15.2). We suggest that Indigenous storytelling provides an excellent medium to support local adaptive capacity and to tailor climate change communication to local contexts, helping to connect global climate change discourses to local realities.

Indigenous Storytelling for ILK Revitalization

Indigenous stories reflect the co-evolutionary dynamics of nature and culture interactions. Through storytelling, elders pass down the social memory of landscape and ecological changes, as informed by Indigenous epistemic traditions (Davidson-Hunt and Berkes, 2003; Fernández-Llamazares et al., 2015a). Oral traditions in various Pacific Island cultures are able to describe ecological events that occurred as long as 700 years ago (Janif et al., 2016). Similarly, it has been shown that the oral traditions of the Klamath Indigenous peoples of North America still pass on the collective memory of the volcanic eruption of Mount Mazama that took place more than 7000 years ago (Deur, 2002; Barber and Barber, 2004). Such collective biocultural memory is increasingly recognized as a source of social-ecological resilience, directly influencing local agency for adaptation to the impacts of global environmental change, including climate change (Sakakibara, 2008; Bali and Kofinas, 2014).

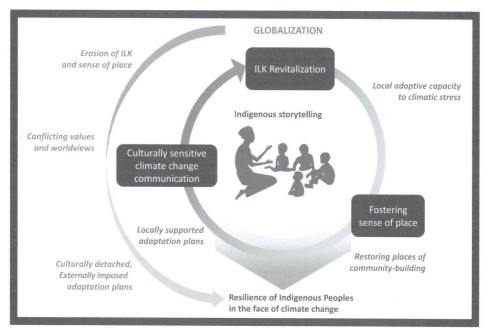

Figure 15.2 Conceptual model describing the role of Indigenous storytelling in climate change adaptation planning. Storytelling is a useful tool for achieving different adaptation goals aimed at fostering social-ecological resilience among Indigenous peoples (blue). Through the promotion of storytelling, such goals can be translated into a series of responses promoting culturally-relevant adaptation (green) while also addressing some of the impacts of globalization upon Indigenous cultures (red). [A black and white version of this figure will appear in some formats. For the colour version, please refer to the plate section.]

Stories, myths, legends and folktales transmit fundamental metaphors that structure the relationship between Indigenous peoples, the environment and the spiritual world (Lawrence and Paige, 2016). Culturally based cosmological explanations are ubiquitous in Indigenous stories, explaining the way in which things are interconnected and promulgating principles that regulate human–nature relations, as well as the role of humans in the natural world (Sakakibara, 2008). Stories of Mother Earth shared by the Indigenous peoples of the South American Andes emphasize the collective cosmo-centric relationships across time between people and nature (e.g., Díaz et al., 2015). Based on the moral guidance of some of these stories, changes in the environment, such as climate change or changes in the availability and/or distribution of natural resources, are often interpreted in spiritual terms and become culturally relevant through their integration into Indigenous peoples' ideologies and stories. Such socio-cultural representations are of utmost importance because they also shape how people might respond to the impending changes. In other terms, it is based on the knowledge transmitted and/or conceptualized through stories that many Indigenous peoples determine whether or not change merits a certain response or adaptation. Thus, stories have direct impacts, not only on how climate change is understood in Indigenous communities but also on how such communities adapt to it (Garteizgogeascoa et al., 2020). As stories partake in creating realities (Blaser, 2014), they play an important role in determining the adoption of management practices that govern and navigate the complex challenges imposed by climate change.

However, as Indigenous communities undergo rapid social, cultural and environmental changes, their traditional knowledge often becomes substantially eroded, thus undermining local adaptive capacity to face climatic changes (e.g., Ford et al., 2006, 2010; Fernández-Llamazares et al., 2015a). In this context, the ability of local and Indigenous peoples to adapt to climate change cannot be considered as unlimited, unless efforts to protect ILK are taken more seriously. Policy frameworks often focus only on maintaining traditional knowledge itself, fixing it in place and time through documentation in repositories or databases. This approach to preserving biocultural heritage leads to the compilation of static practices in frozen forms. There is thus a need for revitalization approaches that target the social-ecological contexts in which ILK is produced, shared and transmitted (Gómez-Baggethun and Reyes-García, 2013; McCarter et al., 2014).

Given that Indigenous storytelling is a central mechanism for the transmission of ILK, there is increasing interest in using Indigenous storytelling in knowledge revitalization efforts. In Vanuatu, a number of school programmes have encouraged the use of storytelling as a way to integrate the intergenerational transmission of knowledge into the school curricula (McCarter et al., 2014). Similarly, Aikenhead (2001) has documented the formation of 'rekindling traditions' projects, including storytelling, for the reconnection of traditional pathways for knowledge transmission between elder and younger generations in Northern Canada. Both approaches emphasize that storytelling is likely to support biocultural conservation goals (Singh et al., 2010; Ryan, 2015). Along these lines, Heras and Tàbara (2014) cite the 'Caravana Cultural' project in Mexico as an example of successfully employing Indigenous storytelling to trigger intergenerational dialogue within Indigenous communities in relation to different environmental pressures, including climate change.

There is also a plethora of examples showing how stories play a key role in teaching Indigenous peoples how to avoid, or respond to, natural hazards and climate risks (Cronin and Cashman, 2008; Nunn, 2009; Fernández-Llamazares and Lepofsky, 2019). Severe weather phenomena, such as hurricanes or storms, often stimulate the development of stories which are encoded within traditional worldviews and beliefs (Walshe and Nunn, 2012). An interesting example comes from the Simeulue Island of Indonesia, where the *smong* oral history enabled the local populations to escape devastation by the 2004 Indian Ocean tsunami (McAdoo et al., 2006). In particular, oral traditions recount the massive tsunami of 1907 and advise running to the hills after tremors lasting more than one minute. Most Simeulue survivors were familiar with these oral stories and what to do in case of an earthquake. Similarly, *kastom* oral traditions in Vanuatu have been shown to contribute to informing effective responses to natural hazards, for instance by avoiding habitation of low-lying areas (Walshe and Nunn, 2012). Comparable stories are found among other Indigenous groups in areas prone to climate risks (McMillan and Hutchinson, 2002; King et al., 2007). There is little doubt that Indigenous stories form the basis of people's understandings of many climatic hazards and ways to respond to them.

Indigenous Storytelling for Fostering a Sense of Place

The concept of 'sense of place' embeds all dimensions of people's interpretations of the environment in emotional, spiritual and cognitive terms (Jorgensen and Stedman, 2006), including attachment to places, local identity and symbolic meaning (Hausmann et al., 2016). Several authors have shown how, globally, this notion can contribute to actions for conservation and climate change adaptation, such as by creating culturally appropriate management interventions (Heller and Zavaleta, 2009) and promoting public support for related policies (Lokhorst et al., 2014; Hausmann et al., 2016). Indeed, such place-based identity is increasingly being emphasized in adaptation discourses among Indigenous peoples (see Greenop, 2009). Given that virtually all Indigenous stories are 'inseparable from places' (Silko, 1981, p. 55), they connect people to places, thereby maintaining their cultural identities and increasing resilience.

Through a detailed and exhaustive examination of Iñupiat storytelling in Alaska, Sakakibara (2008) shows that Indigenous stories have become crucial for reconnecting local communities with their homelands. Iñupiat stories reveal culturally manifest understandings of climate change, such as through tales of the supernatural, which help people to cope with an unpredictable future while maintaining their connection to disappearing lands. With the environment becoming more unpredictable, these histories have also shifted to accommodate new understandings of climate change as a response to a changing spiritual landscape, gradually incorporating new elements such as the loss of sacred ceremonial places to the seas, thawing of ice cellars in the permafrost, eroding homelands and spirit beings obliged to migrate to escape sea level rise. Such stories stress the restorative benefits of Arctic landscapes as places with a number of spiritual dimensions. In other terms, storytelling renews the kinship between humans and the land and can thus be considered a critical form of cultural adaptation, producing and fostering a collective sense of place. As Sakakibara perceptively states, 'as the rising sea erodes ancestral lands, storytelling weaves old and new homes into a viable place of cultural survival.' (Sakakibara, 2008, p. 461).

With millions of Indigenous peoples expected to be relocated as a consequence of climate change, storytelling could also potentially contribute to diminishing the burden of psychological distress and mental illness caused by relocation policies. As a case in point, Athayde et al. (2002) describe how the promotion of oral traditions in the intercultural schools of the Xingu Indigenous Park in the Brazilian Amazonia has supported the transmission of the cultural identity of the Kaiabi peoples after a number of resettlements into new territories. Along these lines, traditional oral narratives of past relocations in the Pacific Islands are considered as valuable adaptation assets that offer practical demonstrations of how community and cultural integrity can survive such disruptions (Janif et al., 2016). In this sense, promoting Indigenous storytelling practices should be viewed as a potential pathway for restoring spaces of communication and community-building. Bali and Kofinas (2014) report how a participatory videography project for documenting and sharing the traditional stories of caribou-user communities in the North American Arctic helped to empower Indigenous communities. The project developed stewardship strategies, fostered community resilience and enhanced local capacity by transmitting invaluable knowledge on how to deal with uncertainty at multiple scales, thus contributing to holistic understandings of climate change. Indigenous stories also enact different ways of telling that can connect ecological, political, spiritual and place-based meanings of climate change in rather unanticipated ways (see Box 15.1).

Box 15.1
Tsimane' storytelling in an age of climate change

The Tsimane' people of the Bolivian Amazon use traditional myths and stories to explain rapid climatic shifts. A number of Tsimane' stories, generally referred to as the eschatological myths, reveal human attitudes towards natural catastrophes that could potentially wipe out humans and other living creatures from the Earth (Huanca, 2008). The use of the term 'tsäqui'' (meaning 'great danger' in the Tsimane' language) is recurrent in myths providing supernatural explanations for climate change. Moreover, interpretations of climate change anchored in local cultural belief systems – angering the spirits by felling too many trees, for example – also seem to be very common in Tsimane' stories (Fernández-Llamazares et al., 2015b). Indeed, the Tsimane' often warn against culturally inappropriate behaviours through traditional stories about forest spirits that need to be revered. The Tsimane' believe that supernatural deities control

Box 15.1 (cont.)

nature and, consequently, they interpret ecological changes as punishment by the spirits in response to disrespectful conduct that transgresses established cultural norms (Fernández-Llamazares et al., 2017). Most Tsimane' elders still follow these traditional norms and taboos regulating natural resource management and attribute climatic changes to the increasing lack of respect towards forest spirits, particularly by younger generations (Fig. 15.3).

Figure 15.3 Elders are essential for the transmission of the Tsimane' cultural identity. Through storytelling, Tsimane' elders weave spaces of intergenerational dialogue around climate change and teach rules, taboos and rituals in relation to natural resource management.
Photo by Isabel Díaz-Reviriego

Indigenous Storytelling for Culturally Sensitive Climate Change Communication

While it is clear that Indigenous storytelling supports ILK transfer mechanisms including biocultural memory, thus contributing to more accurate perceptions of social-ecological changes, it is perhaps less explored how it could help to assimilate external sources of climate change information. In an article commenting on the Encyclical Letter of Pope Francis '*Laudato si':* Our Care for Our Common Home', Hulme (2015) argued that it is essential to create powerful and meaningful stories that make sense of climate change according to motivational cultures and religious beliefs in order to encourage people to adapt to the impending changes. Yet, adaptation agencies, non-governmental organizations and practitioners often fail to share messages in salient, legitimate and credible ways, particularly when it comes to engaging Indigenous peoples. In view of this, as oral traditions are likely to be more influential than externally imposed messages (Janif et al., 2016), current framings of climate change communication should probably be rethought. Nonetheless, the importance of Indigenous storytelling for promoting adaptation goals has been under-recognized in mainstream development approaches.

A handful of examples shows, however, the strong potential of Indigenous stories to inform adaptive management of natural resources. We contend that greater attention to Indigenous ontologies, as transmitted in Indigenous stories, can contribute to making climate change adaptation and mitigation programmes more culturally sensitive, facilitating intercultural discussion and the bridging of worldviews. Herman (2016) calls for more holistic communication than just that based on scientific facts and stresses the role of storytelling as a means of promoting cultural discourses that support behavioural changes towards more sustainable directions. Collaborative planning would then benefit from approaching Indigenous stories as 'ontological assertions' rather than dismissing them as myths (Watson and Huntington, 2008, p. 269).

Through recording Fijian oral traditions about past environmental changes, Janif et al. (2016) showed that traditions such as narratives on how to recognize the approach of a tropical cyclone might be useful in developing adaptive strategies, particularly in areas where access to radio, television or other mass media is still limited. Most Fijians considered storytelling as the most culturally valid method for communicating local interpretations and knowledge of climate change.

Government agencies relying on oral narratives as a means to design participatory adaptation planning are likely to develop initiatives that are more finely tuned to the views, needs and priorities of local communities, thus finding more receptive audiences (Mercer et al., 2007). Incorporating the cultural values and belief systems of local communities has generally met with significant success through high rates of local support. In this context, Indigenous storytelling has the potential to activate learning institutions, helping to connect local cultural values to global-scale policies and stimulating grassroots action towards locally relevant climate change adaptation planning.

Conclusion

Indigenous stories are well-positioned to offer fresh insights and inspirations relating to what it means to be human in an age of climate change. First, storytelling is likely to strengthen the community bonds that support ILK transmission for adaptation. Second, by engaging with creative imagination and complementing knowledge with feeling, emotions, values and beliefs, Indigenous stories allow the fostering of a sense of place, thus helping to frame new locally based cultural discourses. Third, we contend that greater consideration of Indigenous storytelling can contribute to making climate change communication and adaptation more acceptable to local communities by facilitating intercultural discussions and the bridging of worldviews. Moreover, stories generally reflect areas of cultural importance for Indigenous peoples, thus offering opportunities for tailoring adaptation strategies to the local contexts where they are meant to be implemented.

Acknowledgements

We warmly thank the Tsimane', Daasanach, Gabra and Maasai Indigenous peoples for all the stories and folktales that they have shared with us through the years which have been our central inspiration for writing this chapter. This research was supported by the Academy of Finland (grant agreement numbers 292765 and 311176) and Nordenskiöld-Samfundet. We also thank J. de la Malla and I. Díaz-Reviriego for photograph credits; J. R. de Pinho, V. Di Biase, V. Reyes-García and J. Terraube, who provided useful comments and suggestions at earlier stages of this study; and M. Roué for her careful editing.

References

Aikenhead, G. 2001. Integrating Western and aboriginal sciences: Cross-cultural science teaching. *Research in Science Education*, 31(3): 337–355. https://doi.org/10.1023/A:1013151709605

Athayde, S., Troncarelli, M. C., Silva, G. M., Würker, E., Ballester, W. C. and Schmidt, M. V. C. 2002. Educação ambiental e conservação da biodiversidade: A experiência dos povos do Parque Indígena Do Xingu. In Bensusan, N. (ed.) *Seria Melhor Mandar Ladrilhar? Biodiversidade: Como, Para Que, Por Que*. Brasilia (Brazil): Editora da Universidade de Brasilia, pp. 175–188.

Bali, A. and Kofinas, G. P. 2014. Voices of the Caribou people: A participatory videography method to document and share local knowledge from the North American human- Rangifer systems. *Ecology and Society*, 19(2): 16. http://dx.doi.org/10.5751/ES-06327-190216

Barber E. W. and Barber P. T. 2004. *When They Severed Earth from Sky: How the Human Mind Shapes Myth*. 1st edition. Princeton, NJ: Princeton University Press.

Blaser, M. 2014. Ontology and indigeneity: On the political ontology of heterogeneous assemblages. *Cultural Geographies*, 21: 49–58. https://doi.org/10.1177/1474474012462534.

Cronin, S. J. and Cashman, K. V. 2008. Volcanic oral traditions in hazard assessment and mitigation. In Gratton, J. and Torrence, R. (eds.) *Living under the Shadow: Cultural Impacts of Volcanic Eruptions*. Oakland, CA: Left Coast Press, pp. 175–202.

Davidson-Hunt, I. and Berkes, F. 2003. Learning as you journey: Anishinaabe perception of social-ecological environments and adaptive learning. *Conservation Ecology*, 8(1): 5. http://dx.doi.org/10.5751/ES-00587-080105

Deur, D. 2002. A most sacred place: The significance of Crater Lake among the Indians of Southern Oregon. *Oregon Historical Quarterly*, 103(1): 18–49.

Díaz, S., Demissew, S., Carabias, J., Joly, C., Lonsdale, M., Ash, N., Larigauderie, A., Adhikari, J. R., Arico, S., Báldi, A., Bartuska, A., Baste, I. A., Bilgin, A., Brondizio, E., Chan, K. M. A., Figueroa, V. E., Duraiappah, A., Fischer, M., Hill, R., Koetz, T., Leadley, P., Lyver, P., Mace, G. M., Martin-Lopez, B., Okumura, M., Pacheco, D., Pascual, U., Pérez, E. S., Reyers, B., Roth, E., Saito, O., Scholes, R. J., Sharma, N., Tallis, H., Thaman, R., Watson, R., Yahara, T., Hamid, Z. A., Akosim, C., Al-Hafedh, Y., Allahverdiyev, R., Amankwah, E., Asah, T. S., Asfaw, Z., Bartus, G., Brooks, A. L., Caillaux, J., Dalle, G., Darnaedi, D., Driver, A., Erpul, G., Escobar-Eyzaguirre, P., Failler, P., Fouda, A. M. M., Fu, B., Gundimeda, H., Hashimoto, S., Homer, F., Lavorel, S., Lichtenstein, G., Mala, W. A., Mandivenyi, W., Matczak, P., Mbizvo, C., Mehrdadi, M., Metzger, J.P., Mikissa, J. B., Moller, H., Mooney, H. A., Mumby, P., Nagendra, H., Nesshover, C., Oteng-Yeboah, A. A., Pataki, G., Roué, M., Rubis, J., Schultz, M., Smith, P., Sumaila, R., Takeuchi, K., Thomas, S., Verma, M., Yeo-Chang, Y. and Zlatanova, D. 2015. The IPBES Conceptual Framework – connecting nature and people. *Current Opinion in Environmental Sustainability*, 14: 1–16.

Eder, D. J. 2007. Bringing Navajo storytelling practices into schools: The importance of maintaining cultural integrity. *Anthropology & Education Quarterly*, 38(3): 278–296. https://doi.org/10.1525/aeq.2007.38.3.278

Fernández-Llamazares, Á. and Lepofsky, D. 2019. Ethnobiology through Song. *Journal of Ethnobiology*, 39(3): 337–353. https://doi.org/10.2993/0278-0771-39.3.337

Fernández-Llamazares, Á., Díaz-Reviriego, I. and Reyes-García, V. 2017. Defaunation through the eyes of the Tsimane'. In Reyes-García, V. and Pyhälä, A. (eds.) *Hunter-Gatherers in a Changing World*. Cham (Switzerland): Springer International Publishing, pp. 77–90.

Fernández-Llamazares, Á., Díaz-Reviriego, I., Luz, A. C., Cabeza, M., Pyhälä, A. and Reyes-García, V. 2015a. Rapid ecosystem change challenges the adaptive capacity of local environmental knowledge. *Global Environmental Change*, 31: 272–284. https://doi.org/10.1016/j.gloenvcha.2015.02.001

Fernández-Llamazares, Á., Méndez-López, M. E., Díaz-Reviriego, I., McBride, M. F., Pyhälä, A., Rosell-Melé, A. and Reyes-García, V. 2015b. Links between media communication and local perceptions of climate change in an Indigenous society. *Climatic Change*, 131(2): 307–320. https://doi.org/10.1007/s10584-015-1381-7

Fienup-Riordan, A. 1986. *When Our Bad Season Comes: A Cultural Account of Subsistence Harvesting and Harvest Disruption on the Yukon Delta*. Anchorage, AK: Alaska Anthropological Association.

Ford, J. D., Smit, B. and Wandel, J., 2006. Vulnerability to climate change in the Arctic: A case study from Arctic Bay, Canada. *Global Environmental Change*, 16: 145–160. https://doi.org/10.1016/j.gloenvcha.2005.11.007

Ford, J. D., Pearce, T., Duerden, F., Furgal, C. and Smit, B. 2010. Climate change policy responses for Canada's Inuit population: The importance of and opportunities for adaptation. *Global Environmental Change*, 20: 177–191. https://doi.org/10.1016/j .gloenvcha.2009.10.008

Garteizgogeascoa M., García-del-Amo, D. and Reyes-García, V. 2020. Using proverbs to study local perceptions of climate change: a case study in Sierra Nevada (Spain). *Regional Environmental Change*. https://doi.org/10.1007/s10113–020- 01646-1.

Gómez-Baggethun, E. and Reyes-García, V., 2013. Reinterpreting change in traditional ecological knowledge. *Human Ecology*, 41(4): 643–647, http://dx.doi.org/10.1007/ s10745–013-9577-9.

Greenop, K. 2009. *Place Meaning, Attachment, and Identity in Indigenous Inala, Queensland*. Acton (Australia): Institute of Aboriginal and Torres Strait Islander Studies.

Hausmann, A., Slotow, R., Burns, J. K. and Di Minin, E. 2016. The ecosystem service of sense of place: Benefits for human well-being and biodiversity conservation. *Environmental Conservation*, 43(2): 117–127. https://doi.org/10.1017/S0376892915000314

Heller, N. E. and Zavaleta, E. S. 2009. Biodiversity management in the face of climate change: A review of 22 years of recommendations. *Biological Conservation*, 142: 14–32. https://doi.org/10.1016/j.biocon.2008.10.006

Heras, M. and Tabàra, J. D. 2014. Let's play transformations! Performative methods for sustainability. *Sustainability Science*, 9: 379–398. http://dx.doi.org/10.1007/ s11625–014-0245-9.

Herman, R. D. K. 2016. Traditional knowledge in a time of crisis: Climate change, culture and communication. *Sustainability Science*, 11: 163–176. http://dx.doi.org/10.1007/ s11625–015-0305-9

Herman-Mercer, N. M., Matkin, E., Laituri, M. J., Toohey, R. C., Massey, M., Elder, K., Schuster, P. F. and Mutter, E. A. 2016. Changing times, changing stories: Generational differences in climate change perspectives from four remote Indigenous communities in subarctic. *Ecology and Society*, 21(3): 28. https://doi .org/10.5751/ES-08463-210328

Hodge, F. S., Pasqua, A., Marquez, C. A. and Geishirt-Cantrell, B. 2002. Utilizing traditional storytelling to promote wellness in American Indian communities. *Journal of Transcultural Nursing*, 13(1): 6–11. https://doi.org/10.1177/ 104365960201300102

Huanca, T. 2008. *Tsimane' Oral Tradition, Landscape and Identity in Tropical Forest*. La Paz: Imprenta Wagui.

Hulme, M. 2015. Finding the message of the Pope's Encyclical. *Environment*, 57: 16–19. https://doi.org/10.1080/00139157.2015.1089139

Iseke, J. and Moore, S. 2011. Community-based Indigenous digital storytelling with elders and youth. *American Indian Culture and Research Journal*, 35: 19–38. https://doi .org/10.17953/aicr.35.4.4588445552858866

Janif, S. Z., Nunn, P. D., Geraghty, P., Aalbersberg, W., Thomas, F. R. and Camilakeba M. 2016. Value of traditional oral narratives in building climate-change resilience: Insights from rural communities in Fiji. *Ecology and Society*, 21(2): 7. http://dx.doi .org/10.5751/ES-08100-210207

Jorgensen, B. S. and Stedman, R. C. 2006. A comparative analysis of predictors of sense of place dimensions: Attachment to, dependence on, and identification with lakeshore properties. *Journal of Environmental Management*, 79: 316–327. https://doi.org/10 .1016/j.jenvman.2005.08.003

Kesavan, P. C. and Swaminathan, M. S. 2006. Managing extreme natural disasters in coastal areas. *Philosophical Transactions of the Royal Society A: Mathematical,*

Physical and Engineering Sciences, 364(1845): 2191–2216. https://doi.org/10.1098/rsta.2006.1822

King, D. N. T., Goff, J. and Skipper, A. 2007. Māori environmental knowledge and natural hazards in Aotearoa-New Zealand. *Journal of the Royal Society of New Zealand*, 37(2): 59–73. https://doi.org/10.1080/03014220709510536

Lawrence, R. L. and Paige, D. S. 2016. What our ancestors knew: Teaching and learning through storytelling. In Nanton, C. R. (ed.) *Tectonic Boundaries: Negotiating Convergent Forces in Adult Education*. San Francisco: Jossey-Bass, pp. 63–72.

Lokhorst, A. M., Hoon, C., le Rutte, R. and de Snoo, G. 2014. There is an I in nature: The crucial role of the self in nature conservation. *Land Use Policy*, 39: 121–126. https://doi.org/10.1016/j.landusepol.2014.03.005

McAdoo, B. G., Dengler, L., Prasetya, G. and Titov, V. 2006. Smong: How an oral history saved thousands on Indonesia's Simeulue Island during the December 2004 and March 2005 tsunamis. *Earthquake Spectra*, 22(S3): S661–S669. https://doi.org/10.1193/1.2204966

McCarter, J., Gavin, M. C., Baereleo, S. and Love, M. 2014. The challenges of maintaining Indigenous ecological knowledge. *Ecology and Society*, 19(3): 39. http://doi.org/10.5751/ES-06741-190339.

McMillan, A. D. and Hutchinson, I. 2002. When the mountain dwarfs danced: Aboriginal traditions of paleoseismic events along the Cascadia subduction zone of Western North America. *Ethnohistory*, 49 (1): 41–68. https://doi.org/10.1215/00141801-49-1-41

Mercer, J., Dominey-Howes, D., Kelman, I. and Lloyd, K. 2007. The potential for combining Indigenous and Western knowledge in reducing vulnerability to environmental hazards in small island developing states. *Environmental Hazards*, 7(4): 245–256. https://doi.org/10.1016/j.envhaz.2006.11.001

Nakashima, D. J., Galloway McLean, K., Thulstrup, H., Ramos-Castillo, A. and Rubis, J. 2012. *Weathering Uncertainty: Traditional Knowledge for Climate Change Assessment and Adaptation*. Paris and Darwin: United Nations Educational, Scientific and Cultural Organization (UNESCO) and United Nations University Traditional Knowledge Initiative.

Nanson, A. 2011. *Words of Re-enchantment*. 1st edition. Stroud: Awen Publications.

Nunn, P. D. 2009. *Vanished Islands and Hidden Continents of the Pacific*. Honolulu, HI: University of Hawai'i Press.

Ryan, J. C. 2015. The virtual and the vegetal: Creating a 'living' biocultural heritage archive through digital storytelling approaches. *Global Media Journal*, 9(1): 1–10.

Sakakibara, C. 2008. 'Our home is drowning': Iñupiat storytelling and climate change in Point Hope, Alaska. *Geographical Review*, 98(4): 456–475. https://doi.org/10.1111/j.1931-0846.2008.tb00312.x

Silko, L. 1981. Language and literature from a Pueblo Indian perspective. In Fiedler, L. A. and Baker, H. A. (eds.) *Opening up the Canon*. Baltimore, MD: John Hopkins University Press, pp. 54–72.

Singh, R. K., Pretty, J. and Pilgrim, S. 2010. Traditional knowledge and biocultural diversity: Learning from tribal communities for sustainable development in northeast India. *Journal of Environmental Planning and Management*, 53: 511–533.

Walshe, R. A. and Nunn, P. D. 2012. Integration of Indigenous knowledge and disaster risk reduction: A case study from Baie Martelli, Pentecost Island, Vanuatu. *International Journal of Disaster Risk Science*, 3(4): 185–194. https://doi.org/10.1007/s13753–012–0019-x

Watson, A. and Huntington, O. H. 2008. They're *here* – I can *feel* them: The epistemic spaces of Indigenous and Western knowledges. *Social & Cultural Geography*, 9(3): 257–281. https://doi.org/10.1080/14649360801990488

16

Indigenous Knowledge and the Coloniality of Reality: Climate Change Otherwise in the Bolivian Andes

ANDERS BURMAN

Introduction

Several studies have shown that Indigenous peoples are among the most vulnerable to the effects of climate change, and attention has been drawn to Indigenous knowledge as a crucial component of climate change adaptation strategies (see e.g., Kronik and Verner, 2010, p. 1). This chapter argues, however, that in order to take Indigenous 'knowledge' seriously – that is, as something other than 'culture' in supposed opposition to 'science' – Indigenous 'realities' need to be taken seriously. This is because knowledge is not produced in an ontological void. Rather, knowledge is produced in relation to notions of 'what there is', that is, in relation to ontological notions concerning the nature of reality and being (Burman, 2016). Moreover, in order not to make a mere instrumentalist use of Indigenous knowledge as, for instance, in the pharmaceutical industry through bioprospecting for active substances in Indigenous pharmacopeia, this paper argues that, along with instrumental outcomes of Indigenous knowledge, the ontologically informed lifeworlds within which this knowledge is generated ought to be acknowledged. This argument is launched not necessarily in order to argue for the recognition of worlds in the plural and 'the complete dissolution of the notion of an objective, universal nature' (Hornborg, 2015, p. 51), but rather for the more modest acknowledgement that the world we all inhabit is not only epistemically diverse but also multiple in an ontological sense.

The epistemic dimension of continuous colonial domination and the existence of epistemic violence as a fundamental part of the asymmetric relations of power that have characterized the world since 1492 have been discussed in a prolific scholarly debate in terms of the 'coloniality of knowledge' (Grosfoguel, 2013). Critical questions have been raised about who is considered a legitimate producer of knowledge and whose knowledge is considered knowledge. For instance, when compared to 'Western' scientific knowledge, Indigenous knowledge is rarely seen as being of equal merit. While much has been said on the matter, an essential dimension is overlooked in the debate on the coloniality of knowledge, especially in relation to climate change and Indigenous knowledge. What is missing is the fundamental discussion about 'what there is' and the mechanisms by which a dominant reality and a dominant understanding of climate change are imposed on other realities and understandings, and how these dynamics are related to global asymmetric relations of power. What is missing, then, is a debate on the coloniality of *reality*: in other words, a discussion concerning the political ontology of climate change. How do we deal with knowledge based on ontological premises other than the ones sanctioned by modernity? That is, how do we deal with 'climate change otherwise'? Acknowledging the value of Indigenous knowledge in climate change adaptation strategies might, to a certain extent, be a way to address the coloniality of knowledge. However, if Indigenous knowledge is acknowledged but Indigenous ontological lifeworlds from within which such knowledge is produced are denied, then we risk ending up reproducing the coloniality of reality, that is, the dynamics by which a certain reality is naturalized and reproduced, while other realities are denied.

Based on many years of ethnographic research with and among Aymara people in the Bolivian Andes, I therefore intend to discuss the following question: How are the partial connections between different ways of producing knowledge, of experiencing and explaining climate change and of experiencing and generating realities turned into spaces of domination and resistance? This is a question dealing with 'what there is', with what kind of agentive actors there are in the world and what kind of intentional beings constitute the relational fields within which knowledge production and political mobilizations take place, and within which climate change is experienced, explained, understood and addressed (see also Burman, 2017). It is, therefore, a question of an ontological nature.

Climate Change in the Andes

Owing to its geographical position and its high altitude, the Andean high plateau is expected to experience serious problems if climate change causes more than a 2°C increase in global temperature. Glaciers will melt; the seasonal rains rising from the Amazon basin will be significantly less abundant and predictable; drought and water shortage will prevail; and the consequences for agricultural production, but also for urban livelihoods, will be severe (Hoffmann and Requena, 2012). Climate change is not, however, only a distant threat for future generations. Among the rural population of the Andes, changing weather patterns are an experiential reality as climate change is currently making itself felt. One rural Aymara leader in his mid-30s explained it to me thus in May 2015,

It was different before. There was a time for sowing, a time for harvesting. Not now. Nowadays it comes later in the year in some places, and earlier in other places. There is climate change, and it worries us. And we discuss it. It shouldn't be raining now, for example. It usually rains in December, January, but now the rains have continued, and there has been hail and frost. [. . .] There has really been a climate change.

There are hailstorms, torrential rains, frost and drought, and above all, there is uncertainty. First, there is uncertainty because people experience seasonal weather conditions that are less foreseeable, as reflected in the following quote of a middle-aged rural Aymara man living in a community on the shores of Lake Titicaca: 'It may rain any time nowadays. There is frost when we don't want it and there is no frost when we need it to make *ch'uño*.[1] The weather is crazy.' Second, there is uncertainty because in people's experiences, traditional indicators for weather forecasting such as the activities, movements, appearances or positions of birds, toads, mammals, insects, plants, clouds, the stars and the moon are less reliable owing to the changing and unpredictable weather conditions (Araujo Cossío, 2012; Tapia Ponce, 2012). As expressed by a rural Aymara man in his 60s, 'Nature doesn't speak to us as before. We used to know what weather there would be. Now nature says one thing and does another thing.' Others argue that this problem is related more to people's incapability of listening and understanding the signs that nature sends than to nature sending inaccurate signs. I asked Carlos Yujra, an Aymara shaman who tragically passed away in 2019 and with whom I worked for many years, about this and he answered, 'People don't know how to listen anymore. Nature speaks just as good as before, but no one listens, no one understands. People don't even know who is the *marani* this year.' Be that as it may, the mention of the *marani* – the Aymara word for the mountain said to govern the weather in a certain year – points to the core topic of this paper: climate change as understood and explained from within ontologically informed lifeworlds other than the ones sanctioned by modernity, that is, climate change otherwise.

For climate scientists and climate activists alike, climate change is primarily about such things as greenhouse gas emissions and CO_2 equivalents, COPs and UNFCCC, mitigation, adaptation and benchmarking, the Keeling curve, CDE and INDCs. In contrast, for Aymara people it is also, and to many primarily, about *achachilas*, *awichas*, *uywiris* and *maranis* who are other-than-human persons with agentive efficacy and intentionality. Whereas *achachilas* are the masculine ancestors embodied as the high mountains and other powerful places (*wak'a*) in the Andean landscape, *awichas* are the feminine ancestors embodied as hills, plains, rivers and lakes. All these place-beings, or earth beings as Marisol de la

Cadena (2015) calls them, are sometimes generically referred to as *uywiri*, which means protector, shepherd or breeder. Among these beings, the *marani* is the authority in charge of governing the weather and is elected annually (*mara* means 'year' and *marani* means something akin to 'place of the year' or 'holder of the year'). These 'place-beings' have a strong sense of morality and therefore they constantly supervise people's whereabouts in order to oversee whether people live a morally sound social life in their local community and whether people fulfil their ritual obligations towards the ancestral beings in the landscape. If people don't behave properly, there are consequences, primarily in the form of extreme weather events (see e.g. Canessa, 2012, pp. 130–131). Carlos Yujra explains,

Every year in August all the *achachilanaka*[2] meet around the highest mountains to choose a new *marani* for the year to come. The *marani* is like a president and the other *achachilas* are like his ministers in a council. They talk about how people have lived their lives, and then they decide who will decide over the weather this year. The weather depends on who is chosen to be *marani*. It can be drought or frost. They can send hail or rain. Some *achachilanaka* are worse than others. It all depends on how the humans have lived their lives and if we have remembered them.

Hence, the perturbation of the weather and the yearly seasons currently experienced by Aymara people is, according to Carlos, caused by people living immoral lives, or in his own words, *jan wali jakaña* ('not good living'), which arouses the resentment of the place-beings.

In Carlos Yujra's world, there are seven major *achachilas* and *awichas* who are part of the governmental 'council'. They all have their particular temperaments and characteristics, which are expressed as distinct weather phenomena the year they are elected *marani*. Every year, in August, Carlos looked to the mountains for signs. When clouds gather around one of the peaks for many days, he would know that mountain had been chosen to be *marani* for the year to come, and he could thus predict the weather. Five of the *achachilas* are described here,

- Illimani rises 6,438 metres above sea level and towers over the city of La Paz and is, according to Carlos Yujra, a reliable and balanced *achachila* who tends to generate likewise stable and predictable weather throughout the year. Illimani is also known for producing good pasturage for the livestock.
- Approximately 40 km to the northwest, watching over the city of El Alto, Qaqajaqi rises 6,088 metres above sea level. The mountain is known as Huayna Potosi on official maps but owing to its imposing white and grey snowcap, Qaqajaqi or Qaqaqi – meaning 'the greyish person' – is the name used by Carlos Yujra and most rural people. Qaqajaqi is a less reliable *marani* than Illimani and tends to send drought and frost which may have devastating consequences for the livelihood of rural communities.

- Some 50 km further to the northwest, Illampu rises 6,368 metres above sea level. Illampu is known for causing a harsh and cold climate, sending snow and frost and whipping up dust from the ground with strong winds.
- To the South, rising 6,542 metres above sea level on the Oruro *altiplano* (high plateau), Sajama is the highest mountain in Bolivia, and according to Carlos Yujra, a very powerful wind-maker. During a year under Sajama's authority, drought is also to be expected.
- With its slopes covered not by snow and ice but by subtropical vegetation, Uchumachi stands on the Eastern slopes of the Andes, 1,677 metres above sea level. According to Carlos, Uchumachi is the 'father and mother of all plants'. During a year when Uchumachi is *marani*, there is no risk of frost or heavy winds, but rather of torrential rains and flooding.

Apart from these five main *achachila* mountains, there are *achachilas* and/or *awichas* who are embodied in places other than mountains and who can be elected *marani*. Lake Titicaca is one such place which is the embodiment of an *awicha* and an *achachila* who together form *Quta Jaqi* or *Qutaqi*. *Quta Jaqi* means 'lake-person', and when elected *marani*, a wet and rainy year is expected and people take precautions to prevent fields and pastures from being flooded. Another powerful being is said to reside on an underwater ridge in Lake Titicaca. She is an *awicha*, not an *achachila*, and her name is *Chuqilla*. When elected *marani*, she brings rain and good harvests. *Chuqilla* is not altogether harmless, though; she can also send devastating lightning.

This conception and explanation of climate change – this climate change otherwise – is, of course, only intelligible in a world that is sentient, knowing and responsive, and in which social relations between persons can 'override the boundaries of humanity as a species' (Ingold, 2000, p. 107). It is in this world, however, that 'Indigenous knowledge' is produced, not primarily as an intellectual activity inside the minds of hermetically sealed human subjects peeking out at a world that is yet to be known, but as knowledge-generating practices and relations between knowledgeable beings of different kinds. In other words, not only do social relations go beyond humanity as a species, but the boundaries of the epistemic community are not drawn around humanity and are permeable to other knowledgeable subjects.

Indigenous Knowledge and Climate Change

The Intergovernmental Panel on Climate Change (IPCC) has, on various occasions (Parry et al., 2007; IPCC, 2010), emphasized the value of Indigenous knowledge for climate change adaptation. In the two volumes of the fifth IPCC assessment report that deals with adaptation and mitigation, there are no fewer than 510 mentions of the word 'Indigenous'. On the one hand, this is indicative of Indigenous peoples' high degree of vulnerability to climate change, while on the other hand, this reveals the great value Indigenous technologies and knowledges are considered to have for climate change adaptation. It is the latter, that is, Indigenous knowledges and technologies in adaptation strategies, which interest me here.

Long before it was conceptualized as 'traditional ecological knowledge', the complexity and the sophistication of Indigenous environmental knowledges made up the ethnographic substance of many anthropological studies. For instance, at a time in history when Indigenous peoples' traditions were generally seen by outsiders as superstitious mumbo-jumbo – a reflection, of course, of the 'coloniality of knowledge' permeating Eurocentric modernity and according to which legitimate knowledge of the world is produced in and by the North – anthropologist Roy Rappaport showed that traditional ritual practices actually tend to be crucial to long-term ecological sustainability and resilience. Rappaport made a distinction between 'cognized models' and 'operational models'. The cognized model is the 'model of the environment conceived by the people who act in it' that 'elicits behaviour that is appropriate to the material situation of the actors' (Rappaport, 1984, pp. 238–239). While it may prove to have adaptive instrumental significance, that is, positive environmental outcomes, the implicit premise here is that the cognized model is built on an inaccurate notion of causality and that it is therefore not a correct account of how human–environmental relations actually work. People may think that their local environments are balanced and their cassava plots flourish because they make offerings to the ancestors; but there are other mechanisms embedded in their ritual practices, taboos and traditions that, on the one hand, have material consequences and, on the other, are forged by material conditions that explain this outcome. By observing and measuring empirical events, practices and material flows, the task of the anthropologist is, therefore, to grasp the true operational mechanisms in the system and to create a more accurate model of causality. This is what Rappaport calls the 'operational model'.

While one could argue that Rappaport was rather arrogant in claiming that it takes a scientist to discover the 'true' function of Indigenous practices, at a time in history when Indigenous peoples were not on the agenda of any conservation organization, except perhaps as a potential threat to wildlife and biodiversity, Rappaport showed that Indigenous peoples' traditional practices were crucial to ecological sustainability, albeit not for the reasons that the people themselves thought. As will be further developed in this chapter, however, dividing the world into operational and cognized models – where the cognized models may be of interest to the ethnographer but true causal relations are held to be found only in the operational models – is problematic since it reproduces the coloniality of both reality and knowledge.

In the Bolivian Andes, the 'traditional ecological knowledge' of Aymara and Quechua people has gained increasing attention as a key component in climate change adaptation strategies, and five areas have been especially emphasized by researchers (see Araujo Cossío, 2012; Tapia Ponce, 2012),

(1) The sociopolitical organization of the community: By dividing the territory into private plots known as *sayañas* for agricultural production and collective lands known as *aynuqas* primarily for herding animals such as sheep, llama, alpaca and cattle but also for agriculture (principally different varieties of potato), a certain degree of food security and resilience for individual households is acquired (Yampara, 2001). This mixed mode of production, both private and collective, moreover involves collective vigilant undertakings to alert the community of incoming extreme weather events, such as hailstorms, flooding and frost that might destroy crops and pasturage.

(2) The simultaneous management of different ecological zones: John Murra (see e.g., 1985) coined the concept 'vertical archipelago' in order to describe pre-colonial Andean systems for controlling land and accessing and distributing resources from different ecological zones. While much of this pre-colonial system has disintegrated, many Andean Indigenous communities still manage ecological zones at different altitudes where different crops are cultivated and different domestic animals are grazed. Consequently, if harvests fail owing to torrential rains and flooding in low-lying zones, the fields at higher altitudes may still render good harvests. Likewise, if fields at higher altitudes are subject to frost during the critical months from January to April, the fields in low-lying zones may still provide food for the year to come. There is, in other words, a geographical distribution of risks. Moreover, the dispersal of the fields significantly reduces the risk of complete crop failure owing to pests and insects.

(3) Different sowing periods: Crops are sown in three different sowing periods known as *nayra sata* ('early sowing', from late July to August), *chika sata* ('mid sowing', in September) and *qhipha sata* ('last sowing' from October to early November) (Flores Apaza, 2005, p. 85). Even if drought or frost affects crops in a certain critical stage of maturation, other crops are spared. There is, in other words, a distribution of risk in time.

(4) High genetic diversity and mixed cropping: Andean agriculture, in general, is characterized by high species richness. Until recently, as shown by Brush (1982; see also Araujo Cossío, 2012), one Andean family could grow up to 50 varieties of potato and up to 200 varieties could be used in one single community. These varieties are not only differently resistant to pests and

climatic conditions but are adapted to different ecosystems and are therefore used according to the vertical archipelago described earlier. Different varieties are also, however, grown in one and the same field, which further reduces the risk of complete crop failure.

(5) Knowledge of indicators for weather forecasting: Andean people have developed a vast knowledge of astronomical, atmospheric, zoological and phytological indicators for predicting the climatic conditions of the agricultural cycle. When to observe what and how to interpret observations are valued skills among Andean rural people. If the constellation known as the Southern Cross (*Chakana*) is clearly observable on 3 May, the coming year will be a cold one; snowfall in July means it will be a rainy year; fog on August mornings indicates that it is time to plough the fields; in August, if the fox's faeces contain small pieces of branches and twigs, the year to come will be harsh with bad harvests; swarming winged ants in November and December forewarn about incoming frost and the interruption of the rains; if the *q'ua* plant (*Satureja boliviana*) presents few flowers in September and October, the potato harvest will yield very little (for a much more comprehensive list, see Tapia Ponce, 2012). The timely observation of indicators such as these forms a wide-ranging, detailed and sophisticated system to understand the weather and to take preventive action against potential disasters caused by extreme weather phenomena or crop failure.

While 'observations and assessments by Indigenous peoples and local communities have remained largely outside the IPCC process' (Nakashima et al., 2012, p. 24), this kind of knowledge, as mentioned previously, is increasingly referred to in policy documents on climate change adaptation in which it is said to offer 'local precision and nuance' and 'valuable in situ information' (Nakashima et al., 2012, pp. 6–25). Owing to the fact that Indigenous peoples are 'socially and culturally distinct from mainstream society' (Nakashima et al., 2012, p. 6), it is likewise argued that Indigenous knowledge is crucial for any adaptation policy that aims to avoid external impositions. Still, while acknowledging Indigenous knowledge and thereby addressing, to some extent at least, the coloniality of knowledge – that is, the epistemological privilege granted to modern science from the seventeenth century onwards, instrumental not only in legitimizing ethnocide but also in suppressing other forms of knowledge – this approach to Indigenous knowledge is not unproblematic.

From a climate justice perspective, instead of interpreting this approach as a thoroughly benign way of forging culturally sensitive adaptation strategies, one could argue that the recent focus on Indigenous knowledge as a key element for climate change adaptation is a way to make a certain category of people, that is, Indigenous peoples, adapt in economically efficient ways to the changes caused, not by themselves, but by the same powerful actors that currently recognize the value of Indigenous knowledge for climate change adaptation. Technology and knowledge are already in place and adaptation strategies can therefore be implemented in, as IPCC (2010, p. 19) articulates it, 'cost-effective ways'. Moreover, it could be argued to be a way of recognizing Indigenous knowledge only insofar as it is useful within the epistemological and ontological frames set by Western science. In this way, Indigenous knowledge is instrumentalized to serve not only local purposes of adaptation but also the purposes of powerful global actors. A critical debate concerning responsibilities and justice is thereby defanged.

Another problem related to the current focus on Indigenous knowledge for climate change adaptation – and this is my main argument in this chapter – is the all-pervading disembeddedness of knowledge from its ontological context. In otherwise valuable policy-related documents, such as Nakashima et al. (2012), the concept 'worldview', as a mere appendage to 'values and attitudes', tends to be as close as one gets to a serious discussion concerning ontologies. 'Knowledge' is discussed as though delinked from notions of 'what there is', as though knowledge was produced in an ontological vacuum. This, I would argue, is a consequence of dividing the world into 'operational models' and 'cognized models', where the cognized models may be of ethnographic interest but accurate causal relations are held to be found only in the operational models. Nakashima et al. (2012, p. 8) argue that '[i]ndigenous knowledge and knowledge-based practices are the foundations of Indigenous resilience'. This is indeed so. However, the foundations of Indigenous knowledge are Indigenous ontologies. A discussion concerning 'knowledge' should therefore include a discussion about what there is to know and how different notions of what there is to know sometimes coalesce and sometimes collide.

In the Andes, the indicators for weather forecasting discussed earlier in the chapter are not mere biophysical phenomena. They are, according to many Aymara people, signs from *Chimpurachachila*; he is a messenger of the *marani* and the governmental council of place-beings, a sender of signs to be interpreted. A discussion of these indicators that does not take such notions into consideration is ontologically disembedded and therefore inadequate.

These two problems – the instrumentalization of Indigenous knowledge and the ontological disembeddedness – are related since the instrumentalization of Indigenous knowledge is facilitated if the ontological contexts within which Indigenous knowledge is created are ignored. Consequently, while addressing, to a certain extent, the coloniality of knowledge, current approaches advocating increased recognition of Indigenous knowledge in relation to climate change tend, nonetheless, to reproduce the coloniality of reality.

Climate Change and the Coloniality of Reality

Anthropologist Tim Ingold (2007) says we live in a weather world. As human beings, we are all immersed in and conditioned by its fluctuations. Weather, then, seemingly unites us. Nevertheless, weather is experienced, understood, explained and conceptualized in different ways by different actors. Moreover, some are more vulnerable to its potential devastating effects than others – a consequence of social inequalities, not of natural processes. Beyond a concrete local weather world, biologist Eugene F. Stoermer claimed in the 1980s (Crutzen and Stoermer, 2000) that we live in the Anthropocene. In the Anthropocene, human activities constitute such a significant global impact on the Earth's climate that it stands on par with, or even surpasses, geological forces. Humanity is *the* agent. As Malm and Hornborg (2014) and others have convincingly argued, though, there is no undifferentiated humanity that has caused climate change; there is no 'global we' that can be held responsible for greenhouse gas emissions and environmental degradation. The world is, rather, a place of asymmetrically distributed burdens, privileges and risks, not one of equal responsibility. Therefore, the Anthropocene narrative, with its assumption of an undifferentiated humanity, conceals material inequalities and differentiated responsibilities and thereby obstructs the theorization of power (Gooch et al., 2019).

While providing a valuable corrective to this lack of attention to power asymmetries and by drawing attention to a specific historic constellation of capitalism, consumer society and pro-growth ideology, the anthropocentrism of the Anthropocene, and thereby its inapplicability in ontological contexts other than the ones sanctioned by modernity, is left unchallenged by this critique. In the Andes, subjective agentivity is not limited to humanity. There are other-than-human persons with the agentive efficacy to cause extreme weather events and to perturb the yearly seasons, as reflected in the following statement by Carlos Yujra,

People think the weather changes just like that, because of nothing. They don't understand that it's our fault. We are unbalanced and that's why the weather is unbalanced. The *uywiris* get angry. But no one listens to me when I explain that this is what they call climate change.

In her quest to 'reculture and particularize' climate discourses, Georgina Endfield (2011) argues that 'popular conceptualizations and discourses of climate, and its manifestations through local weather, have been replaced by a global, and mainly, scientific metanarrative' (p. 161). In such a powerful metanarrative based on a modernist ontology and shared by proponents and critics of the Anthropocene narrative alike, there would seem to be no role to play for *uywiris* and other place-beings other than that of cultural bric-a-brac. I would argue, though, that the 'ontological opening' identified by Marisol de la Cadena (2014) provides a critical space for such place-beings also in a warming world and that ontologically disobedient actors, such as Carlos Yujra, who carve out counterhegemonic onto-logical spaces, have a critical role to play in addressing climate change and in mobilizing militancy.

The so-called ontological turn has received its fair share of criticism for obstructing the theorization of material inequalities and power. Indeed, critics ask: If the notion of one objective biophysical nature is substituted by notions of a pluriverse (Strathern, 2004) and if the chain of causality behind climate change as explained by science is no more and no less real than any other understanding (ranging from Indigenous understandings to corporate-funded climate scepticism) of the dynamics at work in the atmosphere, how do we articulate a criticism of fossil-fuel led capitalism? Indeed, a relevant question. However, collapsing any understanding of the climate that is not exhausted by modern scientific understandings of causality into climate scepticism is problematic. Indigenous people are not climate sceptics. Carlos Yujra was no climate sceptic and climate change – or *pacha usu*, 'the illness of cosmos' as he would call it in the Aymara language – was to him an experiential reality, not a discourse, narrative or a story. Indigenous people are the first to suffer from the climate change caused by others. In that case, would taking historically subalternized and denied lifeworlds – what I call '*damnés* realities' – seriously in an ontological sense necessarily imply a politically defanged relativism? I think not. Indigenous people are able to recognize more than one chain of causality; they interlace scientific chains of causality and Indigenous, ontologically informed knowledge and experiences into intricate webs of causality (Burman, 2017). To Carlos Yujra, burning fossil fuels was indeed part of *jan wali jakaña* (bad living) and to 'throw toxics out in nature' would make the *uywiris* angry. Humans are by no means acquitted from responsibility in Indigenous understandings of climate change, and power asymmetries are acknowledged and challenged therein. Moreover, when Carlos interweaved different ways of producing knowledge and explaining climate change, he did not reduce one to the other; rather, on the one hand, he carved out spaces for subalternized lifeworlds and denied place-beings within hegemonic structures, and, on the other, he conditionally incorporated, and thereby indigenized, power-infused concepts such as climate change and CO_2 into his own lifeworld.

A scholarly interest in ontological queries, then, does not necessarily imply the dissolution of a biophysical nature or the recognition of a multiplicity of hermetically sealed and discrete worlds; the ontological multiplicity of the world could rather be understood in terms of ontologically informed lifeworlds, that is, worlds as meaningful, experiential and knowable to human beings, worlds from within which power asymmetries are challenged and climate change is understood, experienced and addressed.

If we want to understand Indigenous knowledge and its value for climate change adaptation strategies, we need to understand and take seriously the worlds within which such knowledge is generated. If Indigenous knowledge is acknowledged but Indigenous lifeworlds are denied ontological weight, we risk ending up reproducing the coloniality of reality, that is, the dynamics by which a certain reality is naturalized and reproduced, while other realities are denied.

This, moreover, has a critical bearing on the climate justice debate, as understood and articulated in predominantly non-Indigenous contexts and idioms. Drawing on Boaventura de Sousa Santos' argument that there is no global social justice without global cognitive justice (Santos et al., 2007, p. ixx), I would claim that there is no global climate justice without global cognitive justice (Burman, 2017). And this cognitive justice involves both ontological justice and epistemological justice, that is, it involves both Indigenous knowledge and Indigenous realities.

Notes

1 The Aymara word for freeze-dried potato, a staple food in the Andes, which can be stored for decades.
2 This is the plural form of *achachila* in the Aymara language. In order to facilitate reading, however, I use this plural form only in quotes. In the text, I use the Spanish or English plural form '-s'.

References

Araujo Cossío, H. 2012. *Manejando el Riesgo Climático de Los Andes: El Caso de Las Comunidades Aymara Quechuas de Chillavi-Ayopaya*. La Paz: PIEB.

Brush, S. 1982. The natural and the human environment of the Central Andes. *Mountain Research and Development*, 2(1): 19–38 https://doi.org/10.2307/3672931

Burman, A. 2016. Damnés realities and ontological disobedience: Notes on the coloniality of reality in higher education in the Bolivian Andes and beyond. In Grosfoguel, R., Velasquez, E. R. and Hernandez, R. D. (eds.) *Decolonizing the Westernized University: Interventions in Philosophy of Education from Within and Without*. Lanham, MD: Rowman & Littlefield/Lexington Books, pp. 71–94.

Burman, A. 2017. The political ontology of climate change: Moral meteorology, climate justice, and the coloniality of reality in the Bolivian Andes. *Journal of Political Ecology*, 24: 921–938. https://doi.org/10.2458/v24i1.20974

Canessa, A. 2012. *Intimate Indigeneities: Race, Sex, and History in the Small Spaces of Andean Life*. Durham and London: Duke University Press.

Crutzen, P. J. and Stoermer, E. F. 2000. The "Anthropocene". *IGBP Newsletter*, 41: 17–18.

de la Cadena, M. 2014. The politics of modern politics meets ethnographies of excess through ontological openings. *Theorizing the Contemporary. Cultural Anthropology* website January 13, 2014. https://culanth.org/fieldsights/the-politics-of-modern-polit ics-meets-ethnographies-of-excess-through-ontological-openings

de la Cadena, M. 2015. *Earth Beings: Ecologies of Practice across Andean Worlds.* Durham, NC: Duke University Press.

Endfield, G. 2011. Reculturing and particularizing climate discourses: Weather, identity, and the work of Gordon Manley. *Osiris*, 26(1): 142–162. https://doi.org/10.1086/661269

Flores Apaza, P. 2005. *El Hombre que Volvió a Nacer: Vida, Saberes y Reflexiones de un Amawt'a de Tiwanaku.* La Paz: AOS, PADEM, COSUDE.

Gooch, P., Burman, A. and Almered Olsson, G. 2019. Natural resource conflicts in the Capitalocene. In Almered Olsson, G. and Gooch, P. (eds.) *Natural Resource Conflicts and Sustainable Development.* London: Earthscan, pp. 11–23.

Grosfoguel, R. 2013. The structure of knowledge in Westernized universities epistemic racism/sexism and the four genocides/epistemicides of the long 16th century. *Human Architecture*, 10(1): 73–90.

Hoffmann, D. and Requena, C. 2012. *Bolivia en un Mundo 4 Grados Más Caliente: Escenarios Sociopolíticos Ante el Cambio Climático Para Los Años 2030 y 2060 en el Altiplano Norte.* La Paz: PIEB & Instituto Boliviano de la Montaña.

Hornborg, A. 2015. The political economy of technofetishism: Agency, Amazonian ontol ogies, and global magic. *Hau: Journal of Ethnographic Theory*, 5(1): 47–69. https:// doi.org/10.14318/hau5.1.003

Ingold, T. 2000. *The Perception of the Environment: Essays in Livelihood, Dwelling and Skill.* London and New York: Routledge.

Ingold, T. 2007. Earth, sky, wind and weather. *Journal of the Royal Anthropological Institute*, Special Issue, S19–S38. https://doi.org/10.1111/j.1467-9655.2007.00401.x

IPCC. 2010. Review of the IPCC Processes and Procedures, report by the InterAcademy Council (IPCC-XXXII/Doc. 7), 32nd Session, Busan, Seoul, 11–14 October 2010.

Kronik, J. and Verner, D. 2010. *Indigenous Peoples and Climate Change in Latin America and the Caribbean.* Washington, DC: The World Bank.

Malm, A. and Hornborg, A. 2014. The geology of mankind? A critique of the Anthropocene narrative. *The Anthropocene Review*, 1(1): 62–69. https://doi.org/10 .1177/2053019613516291

Murra, J. 1985. "El Archipielago Vertical" revisited. In Masuda, S., Shimada, I. and Morris, C. (eds.) *Andean Ecology and Civilization.* Tokyo: University of Tokyo Press, pp. 3–13.

Nakashima, D. J., Galloway McLean, K., Thulstrup, H. D., Ramos Castillo, A. and Rubis, J. T. 2012. *Weathering Uncertainty: Traditional Knowledge for Climate Change Assessment and Adaptation.* Paris and Durban: UNESCO & UNU.

Parry, M. L., Canziani, O. F., Palutikof, J. P., van der Linden, P. J. and Hanson, C. E. (eds.) 2007. *Contribution of Working Group II to the Fourth Assessment Report of the Intergovernmental Panel on Climate Change.* Cambridge and New York: Cambridge University Press.

Rappaport, R. A. 1984. *Pigs for the Ancestors.* 2nd edition. New Haven, CT: Yale University Press.

Santos, B. de Sousa, Arriscado Nunes, J. and Meneses, M. P. 2007. Opening up the canon of knowledge and recognition of difference. In Santos, B. de Sousa (ed.) *Another Knowledge Is Possible: Beyond Northern Epistemologies.* London: Verso, pp. xix–xii.

Strathern, M. 2004. *Partial Connections*. Walnut Creek, CA: Altamira.

Tapia Ponce, N. 2012. *Indicadores del Tiempo y la Predicción Climática: Estrategias Agroecológicas Campesinas Para la Adaptación al Cambio Climático en la Puna Cochabambina*. La Paz: PIEB.

Yampara, S. 2001. *El Ayllu y la Territorialidad en los Andes: Una Aproximación a Chambi Grande*. El Alto: UPEA, CADA, INTI-Andino.

Epilogue

17

Negotiating Co-production: Climbing the Learning Curve

IGOR KRUPNIK

In today's Internet-driven era, it takes less than a second to learn that the term 'co-production of knowledge' is hugely popular across many domains. An advanced Google search generates some 688,000 results (as of 1 January 2021) and counting. Naturally, not all these myriad entries refer to the forms of co-production which are the focus of this book, namely those involving scientists, government and environmental agencies, and Indigenous stakeholders. An even smaller segment relates to this book's subtitle, *Global Environmental Change*. Nonetheless, the co-production of knowledge with Indigenous communities is a rapidly expanding field, as we know from science literature, media and personal experience. At times explicitly labelled as such but more often not, it has produced an impressive library of books and papers, to which several authors of this volume have made important contributions (Krupnik and Jolly, 2002; Roué and Nakashima, 2002; Huntington et al., 2004, 2005; Forbes and Stammler, 2009; Salick and Ross, 2009; Nakashima et al., 2012, 2018; Eicken et al., 2014; Roué and Molnár, 2017).

How, then, is this book *different* from so many of its peer publications and why might it stand out among them? My response here is two-fold. First, this book reveals that the incidence of knowledge co-production with Indigenous communities varies widely and is unevenly distributed, both geographically and among major fields within the science of global environmental change. Second, collectively the stories presented in the book offer a new vision of the co-production of knowledge as a *negotiated process*, something that not all its proponents and even its ardent practitioners have fully recognized.

'Genealogy' of Co-production

The conception of the term 'co-production' (coproduction) in the late 1970s is commonly attributed to the political economist Elinor Ostrom (see Roué and

Nakashima, Chapter 1) . Initially, it was devoid of any linkage to Indigenous and local knowledge. It was not associated with the field of environmental science until the late 1980s, until the rise of the 'co-management' approach in fisheries and the use of wildlife resources (Osherenko, 1988), and even a full decade later, as anthropologists and cultural geographers started partnering with Indigenous communities to document local observations and knowledge of climate change (McDonald et al., 1997; Thorpe et al., 2001; Krupnik and Jolly, 2002; Oozeva et al., 2004; Huntington et al., 2005; Lemos and Morehouse, 2005; Crate and Nuttall, 2009). Studies of small-scale subsistence fisheries in the Pacific date further back into the 1970s (Johannes, 1978, 1981; see Roué and Nakashima, Chapter 1).

Although the conceptual toolkit used in this book and the practice of partnering with Indigenous experts and communities for the goal of seeking new knowledge may appear to be rather recent, this is not the case. Indigenous experts have always been indispensable to scientists and, historically, to explorers navigating uncharted waters, crossing inhospitable icy seas and surviving in harsh alien climates. Generations of invaluable environmental and anthropological records, such as dictionaries of Indigenous languages; lists of aboriginal place names; terms for plant, bird and fish species or types of sea ice – all duly documented by scientists since the 1800s – could not have been collected without partnering with local knowledge holders. These interactions, however, would not stand up to today's ethical test of 'co-production'. Even when the parties assumed that they were sharing knowledge and acting in good faith, their interests and understanding of the process were often worlds apart. The power relations between the two types of knowledge – of which one was to be 'extracted' or 'brought in line' with science – were, in general, asymmetrical, whether in the course of the colonial drive for territorial expansion and resource extraction from Indigenous lands or today's international agency mandate for 'sustainable development'.

Hence, the first basic feature of 'co-production' as a scholarly term and as this book's guiding approach is that it requires something *special*, besides merely joining forces in the documentation of someone else's lands, knowledge and languages. One critical condition is a shared vision of *what* is being documented (see Fillipe et al., 2017). It also requires establishing a certain set of ethical principles and a more equitable and balanced interaction between Indigenous knowledge holders and those who work with them or aspire to do so.

The most fundamental question that has challenged and eventually transformed the field of co-production, however, is *for whom* the new knowledge is being generated. Surprisingly, it may have many answers and the best intellectual framework to address its complexity comes from Indigenous people themselves. Almost a decade ago, Gunn-Britt Retter, head of the Arctic and Environmental

Unit of the Sámi Council (www.saamicouncil.net/en/arctic-and-environmental-unit) introduced a valuable typology of environmental scholarship in its relation to the Arctic lands and its peoples. Speaking at the international conference, 'From Knowledge to Action' (Montreal, 2012), she proposed to distinguish among 'knowledge for industry', aimed at collecting data for the optimal exploitation of Arctic resources; 'knowledge for science' that benefits the advancement of academic research and scholarly understanding of polar lands and oceans; and 'knowledge for home', generated for the sustainable future of Indigenous peoples' homelands and cultures. It is the latter form, 'knowledge for home', that most closely aligns with the purpose of various efforts in co-production presented in this book.

In order to establish this intellectual space for rethinking, it is important for knowledge co-production to be viewed as a special path different from the previous agenda of documenting Indigenous people's knowledge – whether practised by international agencies, academic scholars, commercial companies or environmental activists. The rapid rise of 'knowledge for home' and its mounting challenge to the once-dominant role of 'knowledge for science' has been one of the most notable developments of the past decade. Many of this volume's authors have experienced this shift in the course of their professional research, even during a particular project (see Krupnik and Bogoslovskaya, 2017).

Indigenous peoples themselves were key drivers in such transitions through their organized and deliberate actions as part of a new agenda of sovereignty, empowerment and decolonization. This has been particularly evident since the publication of the IPCC AR5 (IPCC, 2014), the ongoing debate on its shortcomings in engaging Indigenous knowledge (see Ford et al., 2016; Nakashima et al., 2018) and the establishment of the Local Communities and Indigenous Peoples' Platform (LCIP Platform) in 2015 (Riedel and Bodle, 2018). In that sense, this book illustrates a new approach in that it advocates the co-production of knowledge as a path towards increasing Indigenous peoples' cultural, political and spiritual resilience, and not only a tool to calibrate climate change assessment, improve management strategies or ameliorate adaptive responses in the face of global environmental change.

Global Inequalities

Despite a large number of recent papers analysing various cases of knowledge co-production with Indigenous communities, few studies address how widely this practice is applied globally. Yet overviews of current research (e.g., Galloway-McLean, 2010; David-Chavez and Gavin, 2018; McElwee et al., 2020) make it clear that the acceptance of knowledge co-produced with Indigenous stakeholders continues to vary widely among geographic regions, nations and scientific disciplines. As the world's Indigenous peoples continue to struggle with past injustices, patronizing attitudes and under-representation, a huge inequality remains in the ways in which their knowledge is being valued (or not) by scientists, agencies, governments and the public.

Some of these differences in acceptance stem from the diverse drivers of knowledge co-production. Multi-national agencies such as FAO, UNESCO or UNEP have historically channelled their attention and resources towards Indigenous practices related to small-scale economies, mainly in tropical forest and semi-desert habitats, such as small-crop cultivation, subsistence use of forest resources, semi-nomadic husbandry, small fisheries, etc. Meanwhile, academic researchers working on the documentation of Indigenous knowledge of climate change introduced biases of their own in that they clearly favoured fieldwork in four geographic settings – the Arctic, low-lying tropical islands, high mountainous areas and deserts (Nakashima et al., 2012; Orlove et al., 2014). However, the recent rise of the Indigenous environmentalist movement has propelled interest in knowledge co-production worldwide. Yet, it also favours better organized (and usually more vocal) Indigenous groups and their organizations, such as the Sámi, the Inuit, or the Nenets of the Russian Arctic (as illustrated by several chapters in this book – see Part 1 as well as Bongo, Chapter 10 and Forbes et al., Chapter 13), along with the Maori and other Indigenous groups in the Pacific. The resulting distribution, as shown in this book, is, again, skewed towards the Arctic, the low-lying tropical islands or tropical regions with strong peasant communities.

This 'skewed geography' evidently has deeper roots. It is the result of past inequalities in resource allocation, or specific interests of certain prime actors or their topics of concern, such as climate change, desertification, economic sustainability or biodiversity conservation. Even if the co-production of knowledge is viewed by so many as an ultimate 'partnering' enterprise, it continues to carry its birthmarks, namely, in the inequality of who chooses whom for what in the process known as 'co-production'. As some recent studies illustrate (David-Chavez and Gavin, 2018; Djenontin and Meadow, 2018; Alexander et al., 2019), it is so often a vision espoused by one side only and the cycle then perpetuates itself.

Negotiating Co-production

If geographic inequities in the co-production of knowledge are a semi-recognized phenomenon, power inequity between Indigenous communities and other actors engaged in the co-production has a longer and better documented history. For decades, Indigenous peoples had to fight an uphill battle to be able to participate in the global environmental change debate. They were excluded from the initial IPCC process started in 1988 that, by definition, was composed of the UN member states only. In the foundational UN Framework Convention on Climate Change report (UNFCCC, 1992), Indigenous peoples were not even mentioned (Macchi, 2008) and their homelands were listed as 'countries' or 'parties', with references to their specific vulnerabilities – 'small island countries', 'forested areas and areas liable to forest decay', 'countries with areas prone to natural disasters', etc. (UNFCCC, 1992, pp. 8–9). The second IPCC Report (AR2 in 1996) made only a passing mention of Indigenous peoples, and in the polar regions only. The change started with the third report (AR3 in 2001; Nakashima et al., 2018) and became more prominent in AR4 (2007) and AR5 (2014), though much remains to be done (Salick and Byg, 2007; Nakashima et al., 2012).

It was not until 1998 that representatives of Indigenous and 'traditional' peoples began participating in the annual UNFCCC Conferences of the Parties (COP) and only since 2001 that they have been acknowledged as a constituency in the climate change negotiations within the UNFCCC. It took another seven years and the adoption of the groundbreaking UN Declaration on the Rights of Indigenous Peoples in September 2007 (UN General Assembly, 2007) to establish the International Indigenous Peoples' Forum on Climate Change (IIPFCC), which created a common political platform for Indigenous organizations and activists fighting to combat global environmental change.

The IIPFCC (www.iipfcc.org) was just one of many voices that articulated the key demands of the world's Indigenous peoples, including equity and respect for their environmental practices and traditional knowledge (Nakashima et al., 2018). One of its pioneering documents (IIPFCC, 2014) argued for Indigenous peoples' rights for full participation in all international climate change agreements, institutions and actions; the recognition of their traditional knowledge; and support for Indigenous community-based monitoring and information systems. These demands have remained high on Indigenous peoples' agenda ever since (as obvious from the list of resolutions, special panels and statements presented at every meeting – see www.iipfcc.org/resources) even if public recognition of the value of Indigenous knowledge has expanded remarkably. Yet the road continues to be 'bumpy' at best, as recalled by Hindou Oumarou in this book (Chapter 8), describing her experiences as Indigenous forum Chair at the UNFCCC COPs where the Platform was discussed and created.

The journey was notably different for another major international environmental treaty, the Convention on Biological Diversity (CBD) that was tabled for signature at the Earth Summit in Rio in June 1992. Here, traditional knowledge of Indigenous peoples was recognized as a valuable contribution since the very beginning and a special working group for its Article 8(j) (' ... respect, preserve and maintain the knowledge, innovations and practices of indigenous and local communities embodying traditional lifestyles … ') was set up early on.

However, the mechanisms of acceptance and inclusion of Indigenous knowledge are still uncertain. Political activism remains a powerful tool in pressing national governments and voicing concerns at international forums, but the practical work has mainly shifted to the new arena of negotiation. This is the context of today's co-production of knowledge debate. Its success will eventually be measured by the newly negotiated space or role for Indigenous knowledge and knowledge-holders in relation to state governments, international agencies and the scientific community engaged in global change research. Therefore, today's knowledge co-production should be viewed through the lens of *Indigenous empowerment*, that is, as a venue to create political and intellectual space for people to speak with authority and power in a balanced and respectful context. It cannot be treated as a one-time or short-term ('one-project') arrangement but rather as a path towards building capacity – including resources, tools and ability – for Indigenous peoples to participate in the co-production of knowledge as equal partners in future collaborative efforts.

A thoughtful synopsis of steps needed to achieve greater equity for Indigenous knowledge was presented recently (Behe et al., 2018; Carlo, 2020, 17–18). This new approach includes several critical elements, such as 'deliberate and intentional choice' for all parties to enter into the co-production process on their terms; 'recognition of Indigenous sovereignty and authority' first and foremost, over peoples' knowledge systems (see United Nations Declaration on the Rights of Indigenous Peoples 2007, Article 31); 'trust and respect' particularly, for the ways in which Indigenous peoples analyse their information, credentials they bring to the table, and the distinctive cultural ways of communication, philosophies and cosmologies at the roots of Indigenous knowledge; and many more. Yet, none are more important than establishing a new set of 'ethical principles and practices' to be placed at the centre of relationships between Indigenous communities, researchers and agencies seeking Indigenous knowledge and recognizing such goals as part of a larger *decolonization* process and as a general endeavour for equity, sovereignty and inclusion.

In an earlier overview (Krupnik et al., 2018, 280–285), we outlined the key contributions that Indigenous peoples have made to the international climate change debate since the 1990s: the focus on 'local scale', the power of self-reliance, the value of innovation and the fight for environmental justice and human rights. As we enter the fourth decade of collective efforts to address the impacts of global change, all these contributions remain highly relevant. The new focus on co-production articulates the emerging framework under which the knowledge systems, environmental monitoring processes and research methodologies of

Indigenous peoples can be treated equally with Western scientific approaches. This book illustrates how this may be done.

Acknowledgements

I am grateful to Douglas Nakashima and Marie Roué for the invitation to write this epilogue for the book and for their many useful editorial comments, and to Gunn-Britt Retter for permission to cite her unpublished presentation made at the 'From Knowledge to Action' conference in 2012.

References

Alexander, S. M., Provencher, J. F., Henri, D. A., Taylor, J. J., Lloren, J. I., Nanayakkara, L., Johnson, J. T. and Cooke, S. J. 2019. Bridging Indigenous and science-based knowledge in coastal and marine research, monitoring, and management in Canada. *Environmental Evidence*, 8: 36.

Behe, C., Daniel, R. and Raymond-Yakoubian, J. 2018. Understanding the Arctic through a Co-production of Knowledge. *ACCAP Webinar*. Fairbanks, AK: Alaska Center for Climate Assessment & Policy, University of Alaska Fairbanks. www.uafaccap.org/event/understanding-the-arctic-through-a-co-production-of-knowledge/

Carlo, N. 2020. *Arctic Observing: Indigenous Peoples' History, Perspectives, and Approaches for Partnership*. Fairbanks, AK: Center for Arctic Policy Studies. www.uaf.edu/caps/our-work/Carlo_Arctic-Observing_Indigenous-Peoples-History_CAPS_5MAR2020.pdf (accessed August 14, 2020).

Crate, S. A. and Nuttall, M. 2009. *Anthropology and Climate Change: From Encounters to Actions*. Walnut Creek, CA: Left Coast Press.

David-Chavez, D. M. and Gavin, M. C. 2018. A global assessment of Indigenous community engagement in climate research. *Environment Research Letters* https://doi.org/10.1088/1748-9326/aaf300

Djenontin, I. N. S. and Meadow, A. M. 2018. The art of co-production of knowledge in environmental sciences and management: Lessons from international practice. *Environmental Management*, 61: 885–903. https://doi.org/10.1007/s00267-018-1028-3

Eicken, H., Kaufman, M., Krupnik, I., Pulsifer, P., Apangalook, L., Apangalook, P., Weyapuk, Jr, W. and Leavitt, J. 2014. A framework and database for community sea ice observations in a changing Arctic: An Alaskan prototype for multiple users. *Polar Geography*, 37(1): 5–27. http://dx.doi.org/10.1080/1088937X.2013.873090

Fillipe, A., Renedo, A. and Marston, C. 2017. The co-production of what? Knowledge, values, and social relations in health care. *PLOS Biology*. https://doi.org/10.1371/journal.pbio.2001403

Forbes, B. C. and Stammler, F. M. 2009. Arctic climate change discourse: The contrasting politics of research agendas in the West and Russia. *Polar Research*, 28: 28–42.

Ford, J. D., Cameron, L., Rubis, J., Maillet, M., Nakashima, D., Willox, A. C. and Pearce, T. 2016. Including Indigenous knowledge and experience in IPCC assessment reports. *Nature Climate Change*, 6: 349–353.

Galloway-McLean, K. 2010. *Advance Guard: Climate Change Impacts, Adaptation, Mitigation and Indigenous Peoples – A Compendium of Case Studies*. Darwin, NT:

United Nations University, Institute for Advanced Studies (UNU-IAS), Traditional Knowledge Initiative, 124 pp.

Huntington, H., Callaghan, T., Fox, S., and Krupnik, I. 2004. Matching Traditional and scientific observations to detect environmental change: A discussion on Arctic terrestrial ecosystems. *Ambio*, 13: 18–23.

Huntington, H., Fox S., Berkes, F., Krupnik, I. 2005. *The Changing Arctic: Indigenous Perspectives. Arctic Climate Impact Assessment Scientific Report*. Cambridge: Cambridge University Press, pp. 61–98.

IIPFCC, 2014. International Indigenous Peoples' Forum on Climate Change. Executive Summary of Indigenous Peoples' Proposal to the UNFCC COP 20 and COP 21. November 2014. Lima, Peru. www.iwgia.org/images/stories/int-processes-eng/UNFCCC/ExecutiveSummaryIPpositionFINAL.pdf

IPCC, 2014: *Climate Change 2014: Synthesis Report. Contribution of Working Groups I, II and III to the Fifth Assessment Report of the Intergovernmental Panel on Climate Change* [Core Writing Team, Pachauri, R. K. and Meyer, L. A. (eds.)]. Geneva: IPCC.

Johannes, R. E. 1978. Traditional marine conservation methods in Oceania and their demise. *Annual Review of Ecology and Systematics*, 9: 349–364.

Johannes, R. E. 1981. *Words of the Lagoon: Fishing and the Marine Law in the Palau District of Micronesia*. Berkeley: University of California Press.

Krupnik, I. and Bogoslovskaya, L. S. 2017. "Our ice, snow and winds": From knowledge integration to co-production in the Russian SIKU project, 2007–2013. In Kasten, E., Roller, K. and Wilbur, J. (eds.) *Oral History Meets Linguistics*. Fürstenberg: Kulturstiftung Sibirien, pp. 31–48.

Krupnik, I. and Jolly, D. (eds.) 2002. *The Earth Is Faster Now: Indigenous Observations of Arctic Environmental Change*. Fairbanks, AK: ARCUS (2nd edition 2010).

Krupnik, I., Rubis, J. T. and Nakashima, D. 2018. Indigenous knowledge for climate change assessment and adaptation: Epilogue. In Nakashima, D., Rubis, J. and Krupnik, I. (eds.) *Indigenous Knowledge for Climate Change Assessment and Adaptation*. Cambridge: Cambridge University Press, pp. 280–290.

Lemos, M. C. and Morehouse, B. J. 2005. The co-production of science and policy in integrated climate assessments. *Global Environmental Change*, 15(1): 57–68.

Macchi, M. 2008. *Indigenous and Traditional Peoples and Climate Change*. Issues Paper. IUCN http://cmsdata.iucn.org/downloads/Indigenous_peoples_climate_change.pdf (accessed April 12, 2015).

McDonald, M., Arragutainaq, L. and Novalinga Z., comps. 1997. *Voices from the Bay: Traditional Ecological Knowledge of Inuit and Cree in the Hudson Bay Bioregion Ottawa: Canadian Arctic Resource Committee, and Sanikiluaq*. NWT: Environmental Committee of Municipality of Sanikiluaq.

McElwee, P., Fernández-Llamazares, A., Aumeeruddy-Thomas, Y., Babai, D., Bates, P., Galvin, K., Guèze, M., Liu, J., Molnár, Z., Ngo, H. T., Reyes-García, V., Chowdhury, R. R., Samakov, A., Shrestha, U. B., Díaz, S. and Brondízio, E. S. 2020. Working with Indigenous and local knowledge (ILK) in large-scale ecological assessments: Reviewing the experience of the IPBES Global Assessment. *Journal of Applied Ecology*, https://doi.org/10.1111/1365-2664.13705

Nakashima, D., Rubis, J. and Krupnik, I. (eds.) 2018. *Indigenous Knowledge for Climate Change Assessment and Adaptation*. Cambridge: Cambridge University Press.

Nakashima, D., Galloway-McLean, K. G., Thulstrup, H., Ramos Castillo, A. and Rubis, J. T. 2012. *Weathering Uncertainty: Traditional Knowledge for Climate Change Assessment and Adaptation*. UNESCO and UNU.

Oozeva, C., Noongwook, C., Noongwook, G., Alowa, C. and Krupnik, I. 2004. *Watching Ice and Weather Our Way/Sikimengllu Eslamengllu Esghapalleghput*. Washington DC: Arctic Studies Center, Savoonga Whaling Captains Association, and Marine Mammal Commission.

Orlove, B., Lazrus, H., Hovelsrud, G. K. and Giannini, A. 2014. Recognitions and responsibilities on the origins and consequences of the uneven attention to climate change around the world. *Current Anthropology*, 55: 249–275.

Osherenko, G. 1988. Can co-management save arctic wildlife? *Environment: Science and Policy for Sustainable Development*, 30(6): 6–34.

Riedel, A. and Bodle, R. 2018. *Local Communities and Indigenous Peoples Platform: Potential Governance Arrangements under the Paris Agreement*. Copenhagen: Nordic Council of Ministers.

Roué, M. and Molnár, Z. 2017. *Knowing Our Lands and Resources: Indigenous and Local Knowledge of Biodiversity and Ecosystem Services in Europe and Central Asia*. Paris: UNESCO.

Roué, M. and Nakashima, D. 2002. Knowledge and foresight: The predictive capacity of traditional knowledge applied to environmental assessment. *International Social Science Journal*, 54(173): 337–347.

Salick, J. and Byg, A. (eds.) 2007. *Indigenous Peoples and Climate Change*. Oxford: Tyndall Centre for Climate Change Research.

Salick, J. and Ross, N. 2009. Traditional peoples and climate change. *Global Environmental Change*, 19: 137–139.

Thorpe, N., Hakognak, N., Eyegetok, S. and Kitikmeot Elders. 2001. *Thunder on the Tundra: Inuit Qaujimajatuqangit of the Bathurst Caribou*. Vancouver: Tuktu and Nogak Project.

UN General Assembly, 2007. 'United Nations Declaration on the Rights of Indigenous Peoples'. Resolution adopted 13 September 2007. www.un.org/development/desa/Indigenouspeoples/declaration-on-the-rights-of-Indigenous-peoples.html (last accessed March 26, 2021).

United Nations Framework Convention on Climate Change (UNFCCC), 9 May 1992. S. Treaty Doc No. 102-38, 1771 U.N.T.S. 107.

Index